M A T E R I A L

an anc ... ...uan text

THE RHIND MATHEMATICAL PAPYRUS

an ancient Egyptian text

GAY ROBINS & CHARLES SHUTE

Published for the Trustees of the British Museum
by British Museum Press

Writing, for one who knows how, is more profitable than all other professions ... See, I am instructing you ... so that you may become one who is trusted by the king, so that you may open treasuries and granaries, so that you may take delivery from the corn-bearing ship at the entrance to the granary, so that on feast-days you may measure out the god's offerings.

Papyrus Lansing, BM 9994

Freeborn children should learn as much of these things as the vast throngs of young in Egypt do with their alphabet. First as regards arithmetic, lessons have been devised there for absolute beginners based on enjoyment and games, distributing apples and garlands so that the same numbers are divided among larger and smaller groups ... The teachers, by applying the rules and practices of arithmetic to play, prepare their pupils for the tasks of marshalling and leading armies and organising military expeditions, managing a household too, and altogether form them into persons more useful to themselves and to others, and a great deal wider awake.

Plato, *Laws* 7, 819

Set in Times and Optima by Filmtype Services Limited, Scarborough, North Yorkshire

Colour reproduction by Colourscan, Singapore
Printed in Hong Kong

Designed by Adrian Hodgkins

CONTENTS

Acknowledgments and Note   6
Preface   7
History of the Papyrus   9
General description   10
Numerals and units of measurement   12
Multiplication and division   16
Addition of fractions and summing to 1 (Problems 7–23)   19
Doubling of unit fractions   22
Division of numbers by 10 (Problems 1–6)   36
Solution of equations and related tables
    (Problems 24–38, 47, 80–1)   37
Unequal distribution of goods and other problems
    (Problems 39–40, 61, 63–5, 67–8)   41
Squaring the circle (Problems 41–3, 48, 50)   44
Rectangles, triangles and pyramids (Problems 44–6,
    49, 51–60)   47
Value, fair exchange and feeding (Problems 62, 66,
    69–78, 82–4)   50
Diversions (Problems 28–9, 79)   54
Conclusions   58
Select bibliography   60
Plates   following 64

Acknowledgments

We wish to thank Mr T. G. H. James, Keeper of Egyptian Antiquities at the British Museum, for his help and for having encouraged us in an undertaking that has proved for us so enjoyable.

We are also grateful to Dr Richard Fazzini, Curator of Egyptian, Classical and Ancient Middle Eastern Art in The Brooklyn Museum, New York, for permission to include a photograph of the fragments now in The Brooklyn Museum (accession no. 37.1784E).

Special thanks are due to Lois and Maurice D. Schwartz of Los Angeles, California for a generous subvention towards the costs of printing this book.

The translations are by the authors and the drawings are by Gay Robins.

Note *by T. G. H. James*

Readers acquainted with earlier publications of the Rhind Mathematical Papyrus will observe that the measurements of the papyrus given in this booklet differ markedly from those previously published. Chace, repeating the figures given by Peet, estimates the original roll at about 543cm in length and 33cm in width, made up of 206cm for BM 10058, 319cm for BM 10057 and 18cm for the gap between the two parts in which the fragments in The Brooklyn Museum are to be fitted. The present measurements for BM 10058 are 199.5cm by 32cm, and for BM 10057, 295.5cm by 32cm; again 18cm are allowed for the gap. The diminution in measurements is due to the conservation work carried out on the papyrus in recent years. After the document was received into the British Museum its two parts were mounted on a backing of card with an unidentified adhesive under some pressure. The deterioration of this card led to splitting and damage (but not loss) to the papyrus. Only during the last ten years have techniques been developed which have allowed the safe removal of long documents from their backing. Using these new techniques conservation staff in the British Museum have been able successfully to detach and remount the two sheets without applying a permanent backing. The removal of the backing and the relaxation of the stresses which had affected the papyrus over more than one hundred years led to the shrinking of the fabric of the sheets. Further reduction in length resulted from the closing of many small gaps where the papyrus and its backing had split. The condition and appearance of the papyrus are now greatly improved.

PREFACE

In this book we have aimed to give a description of the history, form and content of the great Rhind Mathematical Papyrus, hereafter abbreviated to RMP, in such a way as to arouse the interest of the not-too-numerate ancient historian, the educated layman and the student of mathematics who is curious about the remote origins of the subject. In order to make the material more accessible to the modern reader we have adopted some modern conventions. The sums, of course, read from left to right, not from right to left as in the original document. The Egyptians did not use operator signs to indicate addition, subtraction, multiplication and division, but made it clear either verbally or through the method of setting out the calculation what arithmetical process was intended. We have introduced the familiar modern symbols, including the use of brackets for multiplication. Fractions are indicated simply by a bar over a number, since the rule was to use unit fractions only — that is, those with a numerator of 1. The only exception is the fraction $\frac{2}{3}$, which had its own symbol and which is conventionally represented by two bars over the figure 3. The items of the RMP are numbered successively in the manner that has become conventional, not distinguishing between true problems and tables. The problems have been grouped into categories, mainly in the order in which they appear, and representative examples, or those with a special interest, have been given more detailed consideration. In places we have ventured to go beyond the workings provided by the ancient scribe to include some speculations concerning the thought processes that may have underlain his formulation of the problems. Our analysis of the text is preceded by a general account of the mathematical procedures for which the RMP is the main source. In addition to the illustrations to geometrical problems we have included some which feature aspects of daily life in ancient Egypt referred to by the scribe.

HISTORY OF THE PAPYRUS

The Rhind Mathematical Papyrus was discovered in the middle of the last century, allegedly in the ruins of a small building close to the mortuary temple of Ramesses II at Thebes. It was bought in Luxor along with other Egyptian antiquities by Alexander Henry Rhind, who for reasons of health was obliged to winter in Egypt during the seasons of 1855–6 and 1856–7. Rhind died on his way home from another visit to Egypt in 1863, and the RMP, together with another mathematical document known as the Leather Roll (BM 10250), was purchased from his executor in two pieces (BM 10057–8) by the British Museum in 1865. The recto of BM 10057 is on permanent display behind glass in the Third Egyptian Room. Some fragments from the region of the break were identified in the Egyptian collection of the New York Historical Society by the British Egyptologist Professor Percy Edward Newberry in 1922. They had been acquired in Luxor by the American dealer Edwin Smith in 1862–3, and were presented to the New York Historical Society by his daughter after her father's death. They are now in The Brooklyn Museum (no. 37.1784E).

Not surprisingly, the RMP has aroused sustained interest amongst Egyptologists and mathematicians. A version of the hieratic text, pirated from loaned British Museum facsimile plates, together with a translation and commentary, was released by the German Egyptologist August Adolf Eisenlohr in 1877. He also gave the problems the numbers generally used today. The official British Museum facsimile had to await publication until 1898. Professor Thomas Eric Peet, archaeologist and occupier of the chair of Egyptology at the University of Liverpool published an admirable transcription, translation and full commentary in 1923. This was followed by a two-volume edition undertaken by Arnold Buffum Chace, Chancellor of Brown University, together with co-workers, which was intended for mathematicians and laymen rather than Egyptologists. It was published in 1927–9 and reprinted in abbreviated form in 1979, and comprised photographs, a reproduction of the hieratic text with a transcription into hieroglyphs, both in black and red, together with literal and free translations and a very complete bibliography. Certain aspects of the RMP and other texts have been treated at length by Professor Richard J. Gillings in his *Mathematics in the Time of the Pharaohs*, first published in 1972 and reprinted in 1982.

Our views, as expressed in the present book, have naturally been influenced by earlier authors. It must be said, however, that the somewhat exiguous clues provided by the ancients concerning their mathematical methods make it inevitable that modern interpretations should contain an element of subjectivity. Because previous authorities are often in disagreement with each other, it follows that not all that we have to say is in complete accord with what has gone before.

GENERAL DESCRIPTION

The RMP in its original state formed a continuous roll consisting of fourteen sheets of papyrus each about 40cm wide and 32cm high, gummed together at their edges; the total length as it survives today is 513cm. A published version of an inaugural lecture delivered in 1947 by Professor Jaroslav Černý at University College London, gives facts and figures for papyrus rolls dating to the Middle Kingdom and to the Hyksos period immediately following, when the RMP was made. Dimensions of approximately 40cm × 32cm were standard for a full-size papyrus sheet at that time. The second most important mathematical papyrus, dating to the Middle Kingdom and so a little earlier than the RMP, is the well-preserved fragment lodged in the Moscow Museum of Fine Arts (no. 4576), which is formed from quarter-height sheets of 8cm. A length of over five metres for the RMP sounds impressive, but it appears from a New Kingdom account that a complete roll consisted of twenty sheets. The rolls had no central spindle, and so could be comfortably grasped in the hand. It was natural for the scribe to hold the roll in his left hand, and with the brush in his right hand to start, leaving a small margin, at the right end of the papyrus and work from right to left. The inside of the roll would be written on first. This is the surface with the horizontal fibres of the papyrus uppermost. It is known as the recto; the opposite surface, which has the vertical fibres uppermost, is the verso.

The recto of the RMP has, on BM 10058, first the title, date and name of the copyist scribe, then the all-important instructions for doubling odd-numbered unit fractions. The smallest fraction considered, namely $\overline{101}$, occurs on the New York fragments that intervene between BM 10058 and BM 10057. The first sixty problems proper of the RMP are on the recto of BM 10057; these, therefore, are the only ones on actual display in the British Museum. They have a special visual interest on account of the line diagrams that accompany the geometrical group of problems. The verso of BM 10058 has, on the first sheet and so behind the title, no. 61, which is a table and rule for multiplying odd-numbered fractions by $\overline{\overline{3}}$. This is followed by nos. 62–84, bringing the purely mathematical part of the RMP to an end. Number 85, also on the verso of BM 10058, has been variously interpreted as a cryptogram or doodle. The verso of BM 10057 is blank apart from two small areas given the numbers 86 and 87. Number 86 is in fact a patch of three strips pasted on in ancient times; they have on them part of a list of accounts from another papyrus. Number 87, halfway along the top, contains a regnal date of an unnamed king. It is thought by some to throw light on the chronology of the end of the Hyksos period and the Egyptian calendar. Because of their irrelevance to Egyptian mathematics, nos. 85–87 will not be further considered here.

Almost every RMP problem has the opening words picked out in red ink, which helps to demarcate the problems one from another. Sometimes red is used to set apart certain numbers from the main calculation, as in the case of common multiples necessary for

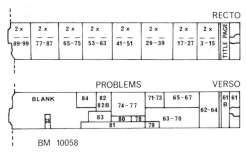

Fig.1a. *Plan of RMP (right), after Chace.*

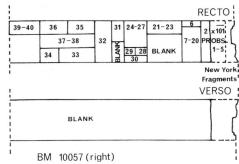

Fig.1b. *Plan of RMP (middle), after Chace.*

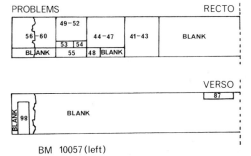

Fig.1c. *Plan of RMP (left),
after Chace.*

the addition of fractions. On the title page it is the title itself that appears in red. It can be rendered: 'Correct method of reckoning, for grasping the meaning of things and knowing everything that is, obscurities ... and all secrets'. Some commentators have held these words to be excessively grandiose, leading to inevitable disappointment with the subsequent contents of the work. Perhaps the high-sounding phrases merely express the pride of the copyist scribe in the methods that he knew how to handle, but may have only partially understood.

The copyist gives his name as Ahmose, and mentions that he is writing in the fourth month of the inundation season in Year 33 of the reign of King Auserre (Apophis). The latter was a king of the Fifteenth Dynasty during the Hyksos or Second Intermediate Period, who flourished around the middle of the sixteenth century BC. Ahmose also records that he is copying earlier work written down in the reign of King Ny-maat-re (Nymare). This was the throne name of Ammenemes III, who was the sixth king of the Twelfth Dynasty and reigned during the second half of the nineteenth century BC. In subsequent pages we shall refer to Ahmose as 'the copyist', and his precursor, or precursors, as 'the scribe'.

NUMERALS AND UNITS OF MEASUREMENT

The Egyptians had a decimal notation. The hieroglyphic script had distinct signs for units, tens, hundreds, etc., the numbers of each being indicated by repetition of the sign. There was no sign for zero and no positional notation, so that the representation of large numbers became extremely cumbersome. The problem was avoided to some extent through the shorthand devices of the hieratic script, which was used for documents. Hieratic was a semi-cursive derivative of hieroglyphs with signs that sometimes reflected their hieroglyphic origins and sometimes, especially in the case of ligatures derived from groups of two or more hieroglyphs, were far removed from them. A striking example is the hieratic sign for 8, where the eight vertical strokes of the hieroglyphic script are replaced by two horizontals.

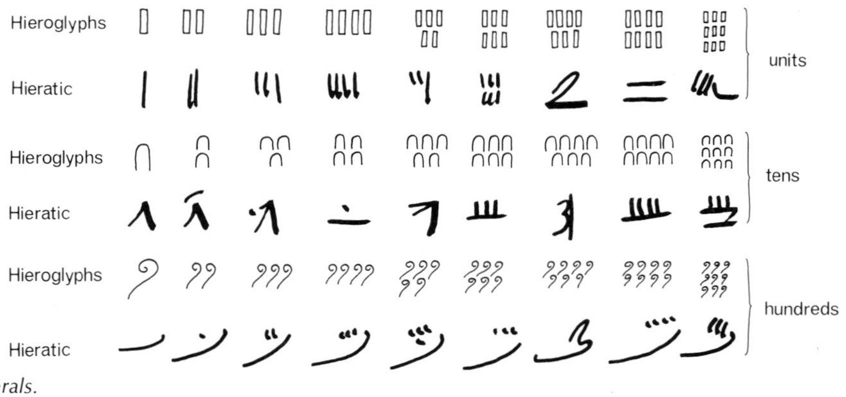

Fig.2. *Numerals.*

Fractions in hieroglyphs were in general indicated by the placement above the number of the sign ⌄, representing pictographically a mouth and phonetically the letter *r*. It is supposed that *r* in this context stood for 'part', as in the modern expression '*n*th part' for the fraction $\frac{1}{n}$. The fraction $\overline{2}$ had a special status with a hieroglyph of its own. The fraction $\overline{3}$ ($\frac{2}{3}$) was represented with two vertical strokes, originally of unequal length, below the *r* (⌄). It is believed that these verticals were intended to indicate the division of a given length in the ratio of 2 to 1. Fractions were represented in hieratic with dots above the number replacing the *r* sign, except in the case of $\overline{\overline{3}}$, $\overline{3}$ and $\overline{4}$, which had signs of their own. The sign for $\overline{4}$ is particularly distinctive to the modern eye, since it forms a cross. The hieratic sign for $\overline{2}$ was similar to the hieroglyph. Sometimes, either by oversight or when it was clear from the context that a fraction was intended, the dot above the number was omitted.

Various RMP problems involve units of measurement belonging to the familiar

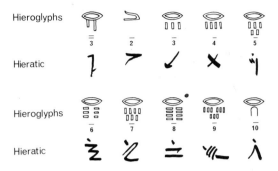

Fig.3. *Fractions.*

categories of length, area and volume. More unusual are the units for slope indicating the degree of flatness of a plane surface such as the face of a pyramid, and for lack of quality indicating a dilution factor in such commodities as bread and beer.

The basic unit of length in the RMP is the royal cubit, corresponding to about 52.5cm and subdivided into 7 palms of 7.5cm, each of which is further divided into 4 fingers of 1.875cm. The names given to these lengths indicate that they are supposed to relate to parts of the forearm (*cubitum* is the Latin for 'elbow'). In fact the royal cubit is longer than the distance from the elbow bone to the middle fingertip in a man with the stature of an ancient Egyptian. It can be shown that in such a person the elbow-to-fingertip distance corresponds instead to the small cubit, which has only 6 palms and so is about 45cm long. This unit does not occur in the RMP, but was probably used for everyday measuring purposes, for which the forearm length would have been particularly convenient. RMP problems are concerned with architectural and land measurements, for which the longer royal cubit was preferred. Land was measured with the help of ropes; a distance of 100 cubits along a rope was called a *khet*.

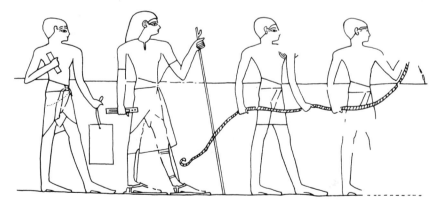

Fig.4. *Measuring a field of corn with a rope, from the tomb of Djeserkaresonb at Thebes.*

The common unit of area was the *setat* or square *khet* for which the Greek word *aroura* is often used as a translation. For smaller areas the *setat* was progressively halved, and special signs were used in the RMP for a half, quarter and eighth of a *setat*, which presumably had special names. In practice a *setat* was also considered to be divisible into narrow strips 1 *khet* long and 1 cubit wide, so that an area of land less than a *setat* consisted of so many fractions of a *setat*, $\overline{2}$, $\overline{4}$ or $\overline{8}$, and so many cubit-strips.

For large areas of land a $10\times$ multiple of the *setat* was used, corresponding to 1000 cubit-strips. In modern terms, the *setat* was about two-thirds of an acre or 0.275 hectare, so that its $10\times$ multiple was about 2.75 hectares.

The common unit of volume, used for measuring amounts of grain or flour, was the *hekat*, approximately equal to 4.8 litres or just over a gallon. For quantities of harvested grain this was obviously too small to be useful, and various multiples of the *hekat* were available. Four times a simple *hekat* amounted to a great or quadruple *hekat* and five times a quadruple *hekat* made a *khar*, equal to two-thirds of a cubic cubit. A common measure for large stores of grain was 100 quadruple *hekat*, equal to 20 *khar*. In modern terms, this was nearly 2 cubic metres or about 53 bushels. For smaller amounts, the *hekat* could be progressively halved to give $\overline{2}, \overline{4}, \overline{8}, \overline{16}, \overline{32}$ and $\overline{64}$ fractions. Subdivision of the unit in this way was convenient because it gave rise to amounts decreasing regularly by half that could be easily visualised. The $\overline{32}$ fraction of a *hekat* was approximately equal to 150cc. A still smaller subdivision was the *ro*, equal to one-tenth of the $\overline{32}$ fraction, and so approximately 15cc.

The $\overline{2}, \overline{4}, \overline{8}, \overline{16}, \overline{32}$ and $\overline{64}$ fractions of the *hekat* are known as Horus-eye fractions, because they were written with distinctive signs that resemble the parts of the eye of the falcon-headed god Horus, known as the *wedjat* eye. In Egyptian mythology the eye of Horus was wounded, wrenched out or eaten by the fearsome god Seth. Later it was restored and made whole, according to Spell 17 in the Book of the Dead, by the ibis-headed god Thoth, the supposed originator of mathematics, who 'did this with his fingers'. The question has been raised whether this phrase referred to 'finger-counting', and whether it could relate in any way to the Horus-eye fractions.

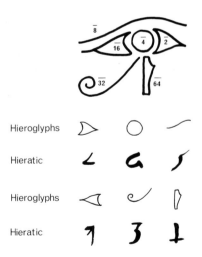

Fig.5. *Horus-eye fractions.*

The $\overline{2}, \overline{4}, \overline{8}, \overline{16}, \overline{32}, \overline{64}$ Horus-eye fractions in modern terminology form a convergent geometrical progression of six terms with the first term equal to the common ratio. RMP no. 79 makes it clear that the Egyptians were familiar with such a series. Today it is elementary to show that the sum of a geometric series to n places is

$$S_n = a\,(1-r^n)/(1-r)$$

where a is the first term and r is the common ratio. In this case, since the first term and the common ratio are equal to $\frac{1}{2}$, the sum to six terms is given by

$$S_6 = \frac{1}{2}[1-(\frac{1}{2})^6]/(1 - \frac{1}{2})$$
$$= 1 - \frac{1}{64}$$

and the sum to infinity is given by

$$S = \frac{1}{2}/(1 - \frac{1}{2})$$
$$= 1$$

showing that the series converges on 1. The Egyptians would have been able to sum the series to six terms by taking a common multiple — that is, multiplying through by 64 (the common multiple) and obtaining the sum 63, showing that the series itself sums to $\frac{63}{64}$. They could also have shown that the sum was short of 1 by $\overline{64}$ by placing $\overline{64}$ at the end, adding it to the preceding term, adding the sum of these to the term before, and so on to the beginning, proving that for the Horus-eye series $S_6 + \overline{64} = 1$. If the Horus – Seth – Thoth story really had a mathematical connotation, it could be that the damaged Horus-eye was magically made whole by the restoration of the missing $\overline{64}$.

The unit of volume for liquids such as beer (also sometimes used for grain) was the *hin*. RMP nos. 80 and 81 reveal that a *hin* was equal to one-tenth of a *hekat*. In practice, liquids were measured by the jugful. Jugs have been found inscribed with their capacity in *hin*, giving on average a value for the *hin* of close to half a litre.

The unit of slope was the *seked*, which measured the lateral displacement in palms for a drop of a royal cubit of 7 palms. Unlike the modern gradient, which depends on the tangent of the angle of inclination of the plane surface and so increases with increasing steepness, the *seked* depended on the cotangent of that angle and so decreased with increasing steepness. The RMP problems involving *seked* deal mainly with the faces of pyramids. By far the commonest *seked* found among actual Old Kingdom pyramids is either $5\frac{1}{2}$ or $5\frac{1}{4}$. The former value appears to be more primitive, since it corresponds to the slope obtained by filling in the steps of the earliest pyramid, the Step Pyramid of Djoser at Saqqara. The faces of the Great Pyramid of Cheops at Giza have an angle of inclination, as measured by surveying, of $51° \ 51'$, corresponding to a *seked* of $5\frac{1}{2}$. For the second pyramid, that of Chephren, the inclination is $53° \ 7'$, corresponding to a *seked* of $5\frac{1}{4}$.

Finally, the unit of measurement registering lack of quality was the *pesu*. It indicated the number of loaves of bread or jugs of beer of standard size that were produced from 1 *hekat* of flour or grain. The greater the *pesu*, the less nutritious the loaf and the weaker the beer. Some such measure was essential in a society where the economy depended on barter, not the use of money. It is interesting that some offering lists record not only the number of loaves and jugs of beer, but also the *pesu*. The *pesu* as applied to bread, or to beer which was made from bread, is sometimes referred to as the 'baking ratio'.

MULTIPLICATION AND DIVISION

The Egyptians multiplied whole numbers in steps. At its most basic, the strategy consisted of successive doubling and adding, which avoided the need to learn multiplication tables. Doubling was carried out by adding a number to itself. With large numbers it was convenient to reduce the number of operations by using 10 as well as 2 as an intermediate multiplier. Thus, if the purpose was to multiply 47 by 33, the sum might be written:

/1	47	or	/1	47
2	94		/10	470
4	188		/20	940
8	376		/2	94
16	752			
/32	1504			
Total 33	1551		Total 33	1551

Dashes were placed against the appropriate intermediate multipliers, as shown, in order to indicate which products should be added to give the required total. This can always be obtained as in the first calculation, since any whole number is expressible as the sum of terms selected from the geometrical progression 1, 2, 4, 8, 16, 32, etc. In effect, the multiplier 33 in the sum 47×33 can be regarded as being partitioned into numbers $(32 + 1)$ taken from this series. In the second calculation, it is partitioned into a different set of numbers $(20 + 10 + 2 + 1)$. The second calculation saves two steps, and the multiplications are simpler.

Multiplication of fractions by fractions is performed in exactly the same way as the multiplication of whole numbers by whole numbers, so that, for instance, $\overline{47} \times \overline{33} = \overline{1551}$. Multiplication of a fraction by a whole number, however, and multiplication of a whole number by a fraction (for the Egyptians these were different processes) both present more of a problem. Consider the examples $\overline{33} \times 47$ and $\overline{47} \times 33$. Theoretically it is possible to obtain the answers to both these sums by successive doubling of the fraction, using the tables for doubled fractions given in the RMP, but the calculations require many steps and much simplification. The preferred alternative was to perform a sum such as $\overline{33} \times 47$ in a special way as a division. To obtain $47 \div 33$, the instruction was given to 'treat 33 so as to obtain 47', and the sum would be written:

/1	33
/$\overline{3}$	11
/$\overline{11}$	3
Total $1 + \overline{3} + \overline{11}$	47

Clearly this is a multiplication in which the divisor 33 of the division sum has become the multiplicand, giving

$$33(1 + \overline{3} + \overline{11}) = 47$$

whence

$$47 \div 33 = \overline{33} \times 47 = 1 + \overline{3} + \overline{11}.$$

The above sum is a simple one because the multiplicand has factors, so it is obvious what fractional multipliers to choose. The situation is more complicated when the divisor is a prime number, as in the case of $\overline{47} \times 33$. It is now required to treat 47 so as to obtain 33. 47 then becomes the multiplicand. It is not immediately apparent what fraction should be used as the first intermediate multiplier, but the Egyptians would have been likely to try first $\bar{\bar{3}}$ or $\overline{2}$. In either case it will be found that the calculation is eased by partitioning 47 into 45 + 2. If $\bar{\bar{3}}$ is chosen as the first intermediate multiplier, the sum goes as follows:

$$
\begin{array}{ll}
\begin{array}{l}
1 \\
/\bar{\bar{3}} \\
/\overline{47} \\
/\overline{94} \\
/\overline{282}
\end{array}
&
\begin{array}{l}
45 + 2 = 47 \\
30 + (1 + \overline{3}) = 31 + \overline{3} \\
\phantom{30 + (1 + \overline{3}) = 31 + } 1 \\
\phantom{30 + (1 + \overline{3}) = 31 + } \overline{2} \\
\phantom{30 + (1 + \overline{3}) = 31 + } 6
\end{array}
\end{array}
$$

Total $\bar{\bar{3}} + \overline{47} + \overline{94} + \overline{282}$ $32 + \overline{2} + \overline{3} + \overline{6} = 33$

The multiplication demonstrates that

$$47(\bar{\bar{3}} + \overline{47} + \overline{94} + \overline{282}) = 33$$

whence

$$33 \div 47 = \overline{47} \times 33 = \bar{\bar{3}} + \overline{47} + \overline{94} + \overline{282}.$$

How was each successive intermediate multiplier in the left-hand column chosen? This is fairly easy to do if the products in the right-hand column are added as the calculation progresses and the amount by which they fall short of the required total, 33, is noted. Thus after the multiplication by $\bar{\bar{3}}$, the shortfall is $33 - (31 + \overline{3}) = 1 + \overline{3}$. After multiplication by $\overline{47}$, it has been reduced to $\bar{\bar{3}}$, and so on.

If, on the other hand, $\overline{2}$ is selected rather than $\bar{\bar{3}}$ as the first intermediate multiplier, the sum becomes

$$
\begin{array}{ll}
\begin{array}{l}
1 \\
/\overline{2} \\
/\overline{5} \\
/\overline{470}
\end{array}
&
\begin{array}{l}
45 + 2 = 47 \\
(22 + \overline{2}) + 1 = 23 + \overline{2} \\
9 + (\overline{5} \times 2) = 9 + \overline{3} + \overline{15} \\
\phantom{9 + (\overline{5} \times 2) = 9 + } \overline{10}
\end{array}
\end{array}
$$

Total $\overline{2} + \overline{5} + \overline{470}$ $32 + \overline{2} + \overline{3} + \overline{10} + \overline{15} = 33$

giving

$$47(\bar{2} + \bar{5} + \overline{470}) = 33$$

or

$$33 \div 47 = \overline{47} \times 33 = \bar{2} + \bar{5} + \overline{470}$$

The multiplication by $\bar{2}$ leaves a shortfall of $33 - (23 + \bar{2})$, which is $9 + \bar{2}$. The next step, multiplication by $\bar{5}$, reduces this to $\overline{10}$, and so on.

Note that the multiplication requires knowledge of the identities $\bar{5} \times 2 = \bar{3} + \overline{15}$ and $\bar{3} + \overline{10} + \overline{15} = \bar{2}$. The Egyptians had this knowledge, as will become clear.

We have produced, then, two possible solutions for $\overline{47} \times 33$ without it being obvious which would have been preferred. The first of the two is the one that can be obtained, in part, by successive doubling of $\overline{47}$. It has a smaller last fractional number (i.e. a larger fraction) than the second, but four fractions instead of three. The choice is finely balanced and would probably have been determined by method rather than merit.

Here is a final example to show how intermediate multipliers are chosen to make the partial products add up to the dividend. Consider the division $100 \div 13$, which occurs in RMP no. 65. The stages of the calculation, although not spelt out, would have gone as follows:

/1	13
/2	26
/4	52
/$\bar{\bar{3}}$	$8 + \bar{\bar{3}}$
$\overline{13}$	1
/$\overline{39}$	$\bar{3}$
Total $7 + \bar{\bar{3}} + \overline{39}$	100

In the above division the divisor 13, which has become the multiplicand, is doubled and redoubled so as to get as close as possible to the dividend, 100. After $7 \times$ (the first three steps), there is a shortfall of 9. After $\bar{\bar{3}} \times$, this is reduced to $\bar{3}$, and after $\overline{39} \times$, it is abolished. The procedure consists of the following steps:

	Partial products	Remainders
13×7	91	$100 - 91 = 9$
$13 \times \bar{\bar{3}}$	$8 + \bar{\bar{3}}$	$9 - (8 + \bar{\bar{3}}) = \bar{3}$
$13 \times \overline{39}$	$\bar{3}$	nil

The Egyptians did not themselves write out divisions in this way, but may have gone through comparable stages in their heads.

ADDITION OF FRACTIONS AND SUMMING TO 1: RMP nos. 7–23

Egyptian arithmetic frequently involved strings of unit fractions, decreasing progressively in size; it was not permitted to repeat a fraction. To make calculation possible it was essential that such fractions could be summed. The addition was achieved, just as it might be today, by taking a common multiple of the fractional numbers involved. Some examples occur in the early section of the RMP that deals specifically with the multiplication of one fractional series by another. In RMP nos. 7, 9–15 the multiplier is $1 + \overline{2} + \overline{4}$ and the multiplicands contain fractions with fractional numbers that are multiples of 7. In nos. 7, 13–15, additions in red have been made to the working, showing that 28 has been taken as the common multiple. In no. 7, the task set is to multiply $\overline{4} + \overline{28}$ by $1 + \overline{2} + \overline{4}$. The calculation consists of the following steps:

$$(\overline{4} + \overline{28})(1 + \overline{2} + \overline{4}) = (\overline{4} + \overline{28}) + (\overline{8} + \overline{56}) + (\overline{16} + \overline{112})$$

$$28\left[(\overline{4} + \overline{28}) + (\overline{8} + \overline{56}) + (\overline{16} + \overline{112})\right] = (7 + 1) + (3 + \overline{2} + \overline{2}) + (1 + \overline{2} + \overline{4} + \overline{4})$$

$$(7 + 1) + (3 + \overline{2} + \overline{2}) + (1 + \overline{2} + \overline{4} + \overline{4}) = 14$$

giving

$$(\overline{4} + \overline{28})(1 + \overline{2} + \overline{4}) = \overline{28} \times 14 = \overline{2}$$

The second step is not explicitly stated, but can be inferred from the fact that $(7 + 1)$, $(3 + \overline{2} + \overline{2})$, $(1 + \overline{2} + \overline{4} + \overline{4})$ is written in red under the partial products of the first step. The equalities $\overline{8} \times 28 = 3 + \overline{2}$ and $\overline{16} \times 28 = 1 + \overline{2} + \overline{4}$ of the second step would have been obtained by multiplication in the usual way:

1	$\overline{8}$	1	$\overline{16}$
2	$\overline{4}$	2	$\overline{8}$
/4	$\overline{2}$	/4	$\overline{4}$
/8	1	/8	$\overline{2}$
/16	2	/16	1
Total 28	$3 + \overline{2}$	Total 28	$1 + \overline{2} + \overline{4}$

It should be mentioned that the workings provided by the scribe in the RMP would have been intended as part of the learning discipline to further the education of the trainee. They do not necessarily indicate how the problem was formulated, or why the scribe selected the particular numbers that he did. The basic structure underlying RMP no. 7 is as follows:

$\bar{4} + \overline{28} = \bar{7} \times 2$ (known from the table of doubled fractions)

$1 + \bar{2} + \bar{4} = \bar{4} \times 7$ (since $7 = 4 + 2 + 1$)

so that

$(\bar{4} + \overline{28})\,(1 + \bar{2} + \bar{4}) = (\bar{7} \times 2)\,(\bar{4} \times 7) = \bar{2}$

The more elaborate working provided by the scribe in the papyrus may have been for educational purposes, to give additional practice in the manipulation of fractions.

In RMP nos. 8, 16–20 the multiplier is $1 + \bar{\bar{3}} + \bar{3}$. The scribe cannot have been unaware that $1 + \bar{\bar{3}} + \bar{3} = 2$. It would have been much simpler just to double the multiplicand, but the object was to demonstrate a method rather than to obtain an answer. The working provided for nos. 19 and 20 shows that the method was again to take a common multiple, this time 18. In RMP no. 19, the instruction is to multiply $\overline{12}$ by $1 + \bar{\bar{3}} + \bar{3}$, and the sum goes as follows:

$\overline{12}(1 + \bar{\bar{3}} + \bar{3}) = \overline{12} + \overline{18} + \overline{36}$

$18(\overline{12} + \overline{18} + \overline{36}) = (1 + \bar{2}) + 1 + \bar{2}$

$(1 + \bar{2}) + 1 + \bar{2} = 3$

giving

$\overline{12}(1 + \bar{\bar{3}} + \bar{3}) = \overline{18} \times 3 = \overline{18}(2 + 1) = \bar{9} + \overline{18} = \bar{6}$

Summing fractional series to 1 was an important part of scribal stock-in-trade, and is exemplified in the RMP 'completion' problems, nos. 21–23. In no. 21, it was requested to complete $\bar{\bar{3}} + \overline{15}$ to make 1. The method, involving 15 as a common multiplier, is shown by the working, which has the following steps:

$15(\bar{\bar{3}} + \overline{15}) = 10 + 1 = 11$

$15 - 11 = 4$

$15(\bar{5} + \overline{15}) = 3 + 1 = 4$

$15[(\bar{\bar{3}} + \overline{15}) + (\bar{5} + \overline{15})] = 11 + 4 = 15$

$(\bar{\bar{3}} + \overline{15}) + (\bar{5} + \overline{15}) = 1$

The answer to the problem, as set, is $\bar{5} + \overline{15}$, but in order to get a series summing to 1, it is necessary to introduce the identity $\overline{15} \times 2 = \overline{10} + \overline{30}$ from the table for doubling fractions to avoid repetition of $\overline{15}$ in the solution. It then follows that the series $\bar{\bar{3}} + \bar{5} + \overline{10} + \overline{30}$ sums to 1, which can be confirmed by taking $\overline{30}$ as a common multiplier. The useful identity $\bar{\bar{3}} = \bar{5} + \overline{10} + \overline{30}$ is a further consequence.

In RMP no. 22, the problem is to complete $\bar{\bar{3}} + \overline{30}$ to 1, not from prior knowledge of the answer to no. 21 but *ab initio* using a similar method, with 30 as the common multiplier. The working is as follows:

$$30(\bar{\bar{3}} + \overline{30}) = 20 + 1 = 21$$

$$30 - 21 = 9$$

$$30(\bar{5} + \overline{10}) = 6 + 3 = 9$$

$$30[(\bar{\bar{3}} + \overline{30}) + (\bar{5} + \overline{10})] = 21 + 9 = 30$$

$$(\bar{\bar{3}} + \overline{30}) + (\bar{5} + \overline{10}) = 1.$$

This time, the information that $\bar{\bar{3}} + \bar{5} + \overline{10} + \overline{30}$ sums to 1 is obtained directly.

In RMP no. 23, it is required to complete the series $\bar{4} + \bar{8} + \overline{10} + \overline{30} + \overline{45}$ to $\bar{\bar{3}}$. Again a similar method is used, with 45 as the common multiplier, yielding $\bar{9} + \overline{40}$ as the answer. A corollary is that

$$\bar{\bar{3}} + \bar{4} + \bar{8} + \bar{9} + \overline{10} + \overline{30} + \overline{40} + \overline{45} = 1$$

It is likely that this problem was made up by the scribe from simpler elements, which can be summarised as follows:

$\bar{3} + \overline{10} + \overline{15} = \bar{3} + \bar{6} = \bar{2}$ $\bar{5} + \overline{20} = \bar{4}$

multiplying by $\bar{3}$ gives   multiplying by $\bar{2}$ gives

$\bar{9} + \overline{30} + \overline{45} = \bar{6}$ $\overline{10} + \overline{40} = \bar{8}$

adding $\bar{3}$ gives   adding $\bar{4} + \bar{8}$ gives

$\bar{3} + \bar{9} + \overline{30} + \overline{45} = \bar{3} + \bar{6} = \bar{2}$ $\bar{4} + \bar{8} + \overline{10} + \overline{40} = \bar{4} + \bar{8} + \bar{8} = \bar{2}$

adding together the two series gives

$$\bar{3} + \bar{4} + \bar{8} + \bar{9} + \overline{10} + \overline{30} + \overline{40} + \overline{45} = 1$$

subtracting $\bar{3}$ gives the identity

$$\bar{4} + \bar{8} + \bar{9} + \overline{10} + \overline{30} + \overline{40} + \overline{45} = \bar{\bar{3}}$$

Various identities, much simpler than this, in which series of fractions are equated to single fractions can be found in the British Museum Leather Roll.

DOUBLING OF UNIT FRACTIONS

Division method

Doubling a fraction when the fractional number is even presents no difficulty: it is readily achieved by halving the fractional number. So the RMP is concerned with the evaluation of $2\bar{n}$ only when n is odd. The importance of the matter is emphasised by its prominent position at the beginning of the papyrus, where all values of n from 3 to 101 are treated. Instructions are given for expressing $2\bar{n}$ as the sum of not more than four different unit fractions of diminishing size, the smallest having a fractional number less than 1000. How this feat was achieved has exercised the minds of Egyptologists and mathematicians for many years.

Anyone attempting to solve this problem must look closely at such information as the RMP provides. Consider, for instance, $n = 17$. The scribe says that 17 is to be treated to get 2, which means performing the division sum $2 \div 17$. The working is as follows:

$\bar{\bar{1}}$	17				
$\bar{3}$	$11 + \bar{3}$		/1	17	
$\bar{3}$	$5 + \bar{\bar{3}}$		/2	34	
$\bar{6}$	$2 + \bar{2} + \bar{3}$				
$/\overline{12}$	$1 + \bar{4} + \bar{6}$	Total 3	51	$\bar{3}$ (in red)	
Remainder	$\bar{3} + \bar{4}$		4	68	$\bar{4}$ (in red)

The first sum shows that $17 \times \overline{12} = 1 + \bar{4} + \bar{6}$ and that $1 + \bar{4} + \bar{6} = 2 - (\bar{3} + \bar{4})$. In other words, the 2 of the doubling process can be dissected into two portions, $(1 + \bar{4} + \bar{6})$ and $(\bar{3} + \bar{4})$. The first portion divided by 17 yields $\overline{12}$. The second sum is intended to show that $\bar{3}$ divided by 17 is equal to $\overline{51}$ and that $\bar{4}$ divided by 17 is equal to $\overline{68}$. For convenience, the calculation is done with whole numbers, giving $17 \times 3 = 51$ and $17 \times 4 = 68$. The red numerals show that the whole numbers are to be turned into fractions. The steps in the reasoning are:

$$2 = (1 + \bar{4} + \bar{6}) + (\bar{3} + \bar{4})$$
$$\overline{17}(1 + \bar{4} + \bar{6}) = \overline{12}$$
$$\overline{17}(\bar{3} + \bar{4}) = \overline{51} + \overline{68}$$
$$\overline{17} \times 2 = \overline{12} + \overline{51} + \overline{68}$$

It should be noted that the first fraction in the expression for $\overline{17} \times 2$ is obtained by successive multiplication of 17 by terms of the fractional series $\bar{\bar{3}}, \bar{3}, \bar{6}$, etc. until a result is reached that is less than 2 and greater than 1, and the remaining fractions consist of $\overline{17}$ multiplied by the fractions required to make this result up to 2.

It is evident that the result obtained for $\overline{17} \times 2$ is not the only one possible. If he wished to be sure that he had obtained the best answer, the scribe would have had to try multiplying 17 by, for instance, members of the $\overline{2}, \overline{4}, \overline{8}$ series until he again reached a result between 2 and 1. The sum would run as follows:

1	17		1	17	
$\overline{2}$	$8 + \overline{2}$		/2	34	$(\overline{2})$
$\overline{4}$	$4 + \overline{4}$		/4	68	$(\overline{4})$
$\overline{8}$	$2 + \overline{8}$		/8	136	$(\overline{8})$
$/\overline{16}$	$1 + \overline{16}$		/16	272	$(\overline{16})$
Remainder	$\overline{2} + \overline{4} + \overline{8} + \overline{16}$				

giving $\overline{17} \times 2 = \overline{16} + \overline{34} + \overline{68} + \overline{136} + \overline{272}$. This result must be rejected because it consists of more than four fractions, but it is possible to get acceptable answers in which, for example, $\overline{10}$ is the first fraction. Problem no. 4 in the RMP gives the information that $\overline{10} \times 7 = \overline{3} + \overline{30}$, and since 17 can be partitioned into $10 + 7$, it follows that $\overline{10} \times 17 = 1 + \overline{3} + \overline{30}$. Now, $1 + \overline{3} + \overline{30}$ can be completed to 2 by adding either $\overline{5} + \overline{10}$ or $\overline{4} + \overline{20}$, because of the identities:

$$\overline{5} + \overline{10} + \overline{30} = \overline{3}$$

$$\overline{4} + \overline{20} + \overline{30} = \overline{3}.$$

It is, therefore, possible to write two expressions for $\overline{17} \times 2$ with $\overline{10}$ as the first fraction, namely

$$\overline{17} \times 2 = \overline{10} + \overline{17}(\overline{5} + \overline{10}) = \overline{10} + \overline{85} + \overline{170}$$

$$\overline{17} \times 2 = \overline{10} + \overline{17}(\overline{4} + \overline{20}) = \overline{10} + \overline{68} + \overline{340}$$

Neither of these is obviously better, in view of the larger numbers in the second and third fractions, than the result obtained more readily by using the $\overline{3}, \overline{\overline{3}}, \overline{6}$ series, unless one attaches special merit to the second expression having all even-numbered fractions, which is advantageous when it comes to doubling.

It so happens that, if one excludes values of n above 5 that are divisible by 3, 5 and 7 (these require separate consideration), all RMP expressions for $2\overline{n}$ with n less than 31 have a first fraction that is a term from the $\overline{3}, \overline{\overline{3}}, \overline{6}$ series, except when $n = 13$. The RMP evaluation of $\overline{13} \times 2$, which uses the $\overline{2}, \overline{4}, \overline{8}$ series, goes as follows:

1	13		1	13	
$\overline{2}$	$6 + \overline{2}$		2	26	
$\overline{4}$	$3 + \overline{4}$		/4	52	$(\overline{4})$
$/\overline{8}$	$1 + \overline{2} + \overline{8}$		/8	104	$(\overline{8})$
Remainder	$\overline{4} + \overline{8}$				

Multiplication of 13 by $\overline{8}$ yields $1 + \overline{2} + \overline{8}$, completion to 2 yields $\overline{4} + \overline{8}$, multiplication of 13 by 4 and 8 yields 52 and 104, so that one can write:

$$\overline{13} \times 2 = \overline{8} + \overline{52} + \overline{104}$$

The alternative of using the $\bar{\bar{3}}, \bar{3}, \bar{6}$ series would have resulted in the following calculation:

	13		1	13	
$\bar{\bar{3}}$	$8 + \bar{\bar{3}}$		$/2$	26	$(\bar{2})$
$\bar{3}$	$4 + \bar{3}$		$/3$	39	$(\bar{3})$
$\bar{6}$	$2 + \bar{6}$		4	52	
$/\overline{12}$	$1 + \overline{12}$		8	104	
Remainder	$\bar{2} + \bar{3} + \overline{12}$		$/12$	156	$(\overline{12})$

giving

$$\overline{13} \times 2 = \overline{12} + \overline{26} + \overline{39} + \overline{156}$$

The series of fractions needed to complete to 2 has an extra term, one of which is odd-numbered, and this leads to a less desirable expression for $\overline{13} \times 2$.

Above $n = 29$, only $n = 37$ and $n = 41$ yield RMP expressions for $2\bar{n}$ with a first fraction from the $\bar{\bar{3}}, \bar{3}, \bar{6}$ series, and none has a first fraction from the $\bar{2}, \bar{4}, \bar{8}$ series. Quite commonly the first fractional number is a multiple of 10. Presumably because of the awkward fractional multiplications involved, the scribe avoided giving full working for higher values of n. Indeed, so much trial and error would have been required to obtain the best solution by division that one is obliged to consider whether a different method could have been employed (see page 28).

Special cases

It is necessary at this point to discuss the doubling of odd unit fractions when n is a multiple of 3, 5 or 7, and the special case when $n = 101$.

RMP expressions for $2\bar{n}$ when n is a multiple of 3 all conform to a single pattern. Those for $2\bar{n}$ when n is a multiple of 5 but not of 3 mostly have an analogous pattern, as do most of those when n is a multiple of 7 but not of 3 or 5, but in both cases there are divergences. It is instructive to compare the preferred expressions to those that would have followed the general pattern.

We will first consider values of n that are multiples of 3. Doubling $\bar{3}$ presents no problem, since $\bar{3} \times 2 = \bar{\bar{3}}$, which becomes the first entry in the RMP table. The first value of n for which $2\bar{n}$ has to be expressed as the sum of unit fractions is $n = 9$. Here a simple calculation shows that $9 \times \bar{6} = 1 + \bar{2}$ and that $9 \times 2 = 18$, whence $\bar{2} = \overline{18} \times 9, \bar{2}$ being the fraction needed to complete $1 + \bar{2}$ to 2. The sum was written in a single column, thus:

$\bar{\bar{3}}$	9	
$\bar{3}$	6	
$\bar{6}$	3	
$/\bar{6}$	$1 + \bar{2}$	
$/2$	18	$\bar{2}$ (in red)

It follows, exactly as for other values of n where $2\bar{n}$ is obtained by the division method, that the first fraction in the expression for $\bar{9} \times 2$ is $\bar{6}$ and the second fraction is $\overline{18}$, giving

$\overline{9} \times 2 = \overline{6} + \overline{18}$

The next fraction to be doubled is $\overline{15}$. If the division method had been used with the terms of the $\overline{\overline{3}}, \overline{3}, \overline{6}$ series as multipliers, the result would have been:

$\frac{1}{3}$	15	1	15
$\overline{3}$	10	/2	30	$(\overline{2})$
$\overline{3}$	5	/4	60	$(\overline{4})$
$\overline{6}$	$2 + \overline{2}$		
$/\overline{12}$	$1 + \overline{4}$		
Remainder	$\overline{2} + \overline{4}$		

giving $\overline{15} \times 2 = \overline{12} + \overline{30} + \overline{60}$. Since $\overline{6} + \overline{30} = \overline{5}$, it follows that $\overline{12} + \overline{60}$ can be simplified to $\overline{10}$, giving $\overline{15} \times 2 = \overline{10} + \overline{30}$, which is the RMP value. We know, however, that the scribe did not do it in this way, since he partitioned 2, not into $(1 + \overline{4}) + (\overline{2} + \overline{4})$, but more simply into $(1 + \overline{2}) + \overline{2}$, writing the sum as follows:

1	15
$/\overline{10}$	$1 + \overline{2}$
$/\overline{30}$	$\overline{2}$

In his head, at least, he may have partitioned 15 into $10 + 5$, which makes the result $15 \times \overline{10} = 1 + \overline{2}$ more obvious.

For all higher values of n where n is a multiple of 3, 2 was similarly partitioned into $(1 + \overline{2}) + \overline{2}$. The first fractional numbers in the expressions for $2\overline{n}$ formed the series 6, 10, 14, etc., increasing in steps of 4. Rather than have to calculate $21 \times \overline{14}$, $27 \times \overline{18}$, $33 \times \overline{22}$, etc., in order to get, in each case, $1 + \overline{2}$, the scribe adopted the admirable device of taking the reciprocal of $1 + \overline{2}$, which is $\overline{\overline{3}}$, and writing $21 \times \overline{\overline{3}} = 14$, $27 \times \overline{\overline{3}} = 18$, $33 \times \overline{\overline{3}} = 22$, etc., with $1 + \overline{2}$ added in red. Multiplying any number by $\overline{\overline{3}}$ was standard practice, and may have been aided by tables. It would, therefore, have been an easy matter to evaluate $2\overline{n}$ when n was a multiple of 3 by taking two-thirds of n and turning it into a fraction (that is, surmounting it with a dot) to get the first term, and doubling n and turning it into a fraction to get the second term. Thus:

$99 \times \overline{\overline{3}} = 66$
$99 \times 2 = 198$
$\overline{99} \times 2 = \overline{66} + \overline{198}$

Clearly, an elegant method has been found for expressing $2\overline{n}$ as the sum of different unit fractions for any value of n that is a multiple of 3. It is not, however, the only one. For instance, the RMP value for $\overline{11} \times 2$ is $\overline{6} + \overline{66}$. Multiplying through by $\overline{9}$ gives $\overline{99} \times 2 = \overline{54} + \overline{594}$, an alternative value, but not to be preferred on account of the large fractional number in the second term.

We will next look at expressions for $2\overline{n}$ in which n is a multiple of 5 but not of 3. These are less straightforward. When $n = 5, 25, 65$ and 85, the 2 of the doubling process is partitioned into $(1 + \overline{3}) + \overline{3}$. Now $1 + \overline{3} = \overline{3} \times 5$, so that the reciprocal of $1 + \overline{3}$ is $\overline{5} \times \overline{3}$. This provides the basis for a rule that, for the above values of n, the first term of the expression for $2\overline{n}$ is got by dividing n by 5, trebling the quotient and turning the result into a fraction. The second term is got by trebling n and turning this result into a fraction. Take $n = 65$ as an example. According to the rule, $65 \times \overline{5} = 13$, $13 \times 3 = 39$, so that the first term of $2\overline{n}$ is $\overline{39}$. Also $65 \times 3 = 195$, so the second term is $\overline{195}$. It seems likely that the scribe would have adopted this simple procedure, although the RMP gives only

$$
\begin{array}{ll}
\quad 1 \qquad\quad 65 \\
/\overline{39} \qquad\quad 1 + \overline{\overline{3}} \\
\quad /3 \quad\quad 195 \qquad\qquad \overline{3} \text{ (in red)}
\end{array}
$$

The sum records that $65 \times \overline{39} = 1 + \overline{\overline{3}}$ and $65 \times \overline{195} = \overline{3}$, whence $65(\overline{39} + \overline{195}) = 1 + \overline{\overline{3}} + \overline{3} = 2$ and so $\overline{65} \times 2 = \overline{39} + \overline{195}$, but there is no indication of the intermediate working.

When $n = 35$, 55 and 95, expressions for $2\overline{n}$ obtainable by the above rule in series with those when $n = 5$, 25, 65 and 85 were rejected in favour of alternatives that the scribe evidently found more attractive:

	rejected	preferred
$n = 35$	$\overline{35} \times 2 = \overline{21} + \overline{105}$	$\overline{35} \times 2 = \overline{30} + \overline{42}$
$n = 55$	$\overline{55} \times 2 = \overline{33} + \overline{165}$	$\overline{55} \times 2 = \overline{30} + \overline{330}$
$n = 95$	$\overline{95} \times 2 = \overline{57} + \overline{285}$	$\overline{95} \times 2 = \overline{60} + \overline{380} + \overline{570}$

These preferred values require some comment. The expression for $\overline{55} \times 2$ is got very easily by division if 55 is partitioned into $30 + 20 + 5$:

$$
\begin{array}{lll}
\quad 1 & 30 + 20 + 5 = 55 & \quad 1 \qquad 55 \\
/\overline{30} & \quad 1 + \overline{\overline{3}} + \overline{6} & /6 \quad 330 \qquad (6) \\
\text{Remainder} & \qquad \overline{6}
\end{array}
$$

giving $\overline{55} \times 2 = \overline{30} + \overline{330}$. Partitioning into $30 + 15 + 10$ would have given the same result, since $\overline{2} + \overline{3} = \overline{3} + \overline{6}$.

The RMP two-term expression for $\overline{35} \times 2$ is unusual in that the fractional number of the second term (42) is not a multiple of n. The first fraction can be got by partitioning 35 into $30 + 5$, whence

$$
\begin{array}{ll}
\quad 1 & 30 + 5 = 35 \\
/\overline{30} & \quad 1 + \overline{6} \\
\text{Remainder} & \quad \overline{\overline{3}} + \overline{6} \text{ (or } \overline{2} + \overline{3})
\end{array}
$$

If $\overline{2} + \overline{3}$ had been chosen as the remainder completing to 2, it would have been possible to write a three-term expression by continuing the division to give:

$$\overline{35} \times 2 = \overline{30} + \overline{70} + \overline{105}$$

But instead, the scribe chose to complete with $\overline{\overline{3}} + \overline{6}$, which does not lend itself to the second part of a conventional division sum. It is likely that he proceeded by equating $\overline{\overline{3}} + \overline{6}$ to $\overline{6} \times 5$ so as to find the reciprocal of $\overline{\overline{3}} + \overline{6}$, which is $\overline{5} \times 6$. Then he would have applied this reciprocal as a multiplier to 35, either partitioning it into $30 + 5$ or factorising it as 7×5, in either case getting 42. The latter alternative is favoured by three accessory numbers, 6 (in red), 7 and 5, that are appended to the RMP working. Since $35(\overline{5} \times 6) = 42$, it follows that

$$\overline{35}(\overline{\overline{3}} + \overline{6}) = \overline{35}(\overline{6} \times 5) = \overline{42}$$

and

$$\overline{35} \times 2 = \overline{30} + \overline{42}$$

A different working for $\overline{35} \times 2$ is given in an early Roman demotic text (BM 10520, see Parker, 1972). There 35 is factorised into 5×7, these factors are added to make 12, which on halving becomes 6. Then 6 applied as a multiplier to the factors of 35 gives 30 and 42. These numbers turned into fractions give $\overline{35} \times 2 = \overline{30} + \overline{42}$. In effect the 2 of the doubling process, instead of being partitioned into $(1 + \overline{6}) + (\overline{\overline{3}} + \overline{6})$ as in the RMP, is taken as being equal to $\overline{6} \times 12$. Then 12 is partitioned into $7 + 5$, so that $2 = (\overline{6} \times 7) + (\overline{6} \times 5)$. Dividing both sides of this equation by $35 = 5 \times 7$ gives $\overline{35} \times 2 = (\overline{6} \times \overline{5}) + (\overline{6} \times \overline{7}) = \overline{30} + \overline{42}$. Notice again the key role played in the calculation by the numbers 6, 7 and 5. The method was almost certainly not known in earlier times. If it had been, it could have yielded for $\overline{55} \times 2$ the attractive expression $\overline{40} + \overline{88}$.

The RMP solution for $\overline{95} \times 2$ is the only example of a three-term expression accepted for $2\overline{n}$ when n is a multiple of 5. Division of 2 by 95 yields $95 \times \overline{60} = 1 + \overline{2} + \overline{12}$ with a remainder of $\overline{4} + \overline{6}$ completing to 2, giving $\overline{95} \times 2 = \overline{60} + \overline{95}(\overline{4} + \overline{6})$. The scribe failed to notice that a simplification was possible, since:

$$\begin{aligned}
\overline{95}(\overline{4} + \overline{6}) &= (\overline{19} \times \overline{5}) \times \overline{2} \times (\overline{2} + \overline{3}) \\
&= \overline{38} \times (\overline{10} + \overline{15}) \\
&= \overline{38} \times \overline{6}
\end{aligned}$$

giving as a two-term expression for $\overline{95} \times 2$:

$$\overline{95} \times 2 = \overline{60} + \overline{228}$$

The same result can be obtained by the method used in the demotic text for $\overline{35} \times 2$, since half the sum of the factors of 95 greater than unity is 12, and these factors when multiplied by 12 give 60 and 228.

We turn now to values of $2\overline{n}$ when n is a multiple of 7 but not of 3 or 5. The cases to be considered are $n = 7, 49, 77$ and 91. In each of these except the last, the 2 of the doubling process is partitioned into $(1 + \overline{2} + \overline{4}) + \overline{4}$. The procedure for obtaining an expression for $2\overline{n}$ is to divide n by 7, quadruple the quotient, and turn the result into a fraction for the first term, and to quadruple n and turn it into a fraction for the second term. This rule follows from $1 + \overline{2} + \overline{4} = \overline{4} \times 7$, so that the reciprocal of $1 + \overline{2} + \overline{4}$ is $\overline{7} \times 4$.

The rule was rejected by the scribe for $n = 91$. In preference to a first fraction of $\overline{52}$, he selected $\overline{70}$, which enables 2 to be partitioned as follows:

$$\frac{1}{7}\Big/\overline{70} \qquad \begin{array}{l} 70 + 14 + 7 = 91 \\ 10 + 2 + 1 = 13 \\ 1 + \overline{5} + \overline{10} \end{array}$$

Remainder $\qquad \bar{\bar{3}} + \overline{30}$

Now $\bar{\bar{3}} + \overline{30} = \overline{30}(20 + 1) = \overline{30} \times 21 = \overline{10} \times 7$, so its reciprocal is $\overline{7} \times 10$. This reciprocal when applied as a multiplier to $91 = 13 \times 7$ gives 130, so that

$$\overline{91} \times 2 = \overline{70} + \overline{130}$$

The RMP value for $\overline{91} \times 2$ can also be obtained by the demotic method described above, since half the sum of the factors of 91 greater than unity is 10, and these factors multiplied by 10 give 70 and 130. The rejected value, given by the general rule, for $2\overline{n}$ when n is a multiple of 7 is:

$$\overline{91} \times 2 = \overline{52} + \overline{364}$$

It remains to consider the RMP expression for $2\overline{n}$ when $n = 101$. Here 2 is equated to $1 + \overline{2} + \overline{3} + \overline{6}$ to give

$$\overline{101} \times 2 = \overline{101} + \overline{202} + \overline{303} + \overline{606}$$

The same procedure could have been used to obtain $2\overline{n}$ for any value of n, but it was avoided except in this instance, no doubt because it would have led to four-term expressions with two odd-numbered fractions. In the case of $n = 101$, however, no alternative exists with not more than 4 fractions and with fractional numbers less than 1000.

Partition of n

The division method involving progressive multiplication of n by terms of the $\bar{\bar{3}}, \overline{3}, \overline{6}$ series proved generally effective (except for $n = 13$) for values of n under 31. With higher values of n it produced expressions for $2\overline{n}$ that were acceptable to the scribe only for $n = 37$ and $n = 41$. With other values of n different multipliers were used; for example, for $n = 31$ and $n = 43$:

$$\frac{1}{}\Big/\overline{20} \qquad \begin{array}{l} 31 \\ 1 + \overline{2} + \overline{20} \end{array} \qquad\qquad \frac{1}{}\Big/\overline{42} \qquad \begin{array}{l} 43 \\ 1 + \overline{42} \end{array}$$

Remainder $\quad \overline{4} + \overline{5}$ $\qquad\qquad$ Remainder $\quad \overline{2} + \overline{3} + \overline{7}$

$2\overline{n} \qquad \overline{20} + \overline{124} + \overline{155}$ $\qquad\qquad$ $2\overline{n} \qquad \overline{42} + \overline{86} + \overline{129} + \overline{301}$

It is interesting to compare these solutions with the best that could have been obtained with the $\bar{\bar{3}}, \overline{3}, \overline{6}$ or $\overline{2}, \overline{4}, \overline{8}$ series, as follows:

1	31	1	43
$\overline{2}$	$15 + \overline{2}$	$\overline{\overline{3}}$	$28 + \overline{\overline{3}}$
$\overline{4}$	$7 + \overline{2} + \overline{4}$	$\overline{3}$	$14 + \overline{3}$
$\overline{8}$	$3 + \overline{2} + \overline{4} + \overline{8}$	$\overline{6}$	$7 + \overline{6}$
$/\overline{16}$	$1 + \overline{2} + \overline{4} + \overline{8} + \overline{16}$	$\overline{12}$	$3 + \overline{2} + \overline{12}$
		$/\overline{24}$	$1 + \overline{2} + \ \overline{4} + \overline{24}$

Remainder $\quad \overline{16}$ $\qquad$ Remainder $\quad \overline{8} + \overline{12}$

$2\overline{n} \qquad \overline{16} + \overline{496} \qquad\qquad 2\overline{n} \qquad \overline{24} + \overline{344} + \overline{516}$

It is by no means self-evident that these variants are inferior in view of the extra terms in the RMP expressions and the odd-numbered fractions, which are awkward to double.

Although the RMP expressions for $\overline{31} \times 2$ and $\overline{43} \times 2$ could have been achieved by division without much difficulty, this is less true for some higher values of n. Take, for instance, $n = 67$. The RMP first fraction for $\overline{67} \times 2$ is $\overline{40}$, and the value given for $67 \times \overline{40}$ is $1 + \overline{2} + \overline{8} + \overline{20}$. No working is provided, but it would have had to go something like this:

$1 \qquad 67$
$\overline{2} \qquad 33 + \overline{2}$
$\overline{4} \qquad 16 + \overline{2} + \overline{4}$
$/\overline{40} \qquad\ 1 + (\overline{2} + \overline{10}) + \overline{20} + \overline{40}$
$\qquad$ (since $\overline{10} \times 6 = \overline{10}(5 + 1) = \overline{2} + \overline{10}$)
$\qquad = 1 + \overline{2} + \overline{8} + \overline{20}$ (since $\overline{10} + \overline{40} = \overline{8}$)

The procedure is laborious and when one takes into account the number of fractional multipliers that would have had to be tried to achieve the most acceptable result there is no doubt that compiling the whole RMP table through straightforward divisions would have taken a dauntingly long time.

In the absence of working, the question of how the calculations were done must be a matter for supposition. But a short-cut could have been achieved by partitioning n. Consider again the case of $n = 67$. An appropriate first fraction needs to have a fractional number that is more than half n and has several factors. Therefore $\overline{40}$ is a good candidate to try. The next step is to partition 67 in such a way that the largest component is 40, and the others, all different, are diminishing factors of 40. The conditions are fulfilled by $67 = 40 + 20 + 5 + 2$, giving $67 \times \overline{40} = 1 + \overline{2} + \overline{8} + \overline{20}$. The remaining fractions for $\overline{67} \times 2$ are got by completing $\overline{2} + \overline{8} + \overline{20}$ to 1 with $\overline{5} + \overline{8}$, and multiplying by $\overline{67}$. The expression for $\overline{67} \times 2$, as given in the RMP, then becomes:

$$\overline{67} \times 2 = \overline{40} + \overline{335} + \overline{536}$$

The advantage of this method is that not only can a value for $2\overline{n}$ be obtained very quickly but so too can alternative solutions. For instance, 67 can also be partitioned into $42 + 21 + 3 + 1$, so that $67 \times \overline{42} = 1 + \overline{2} + \overline{14} + \overline{42}$. Then $\overline{2} + \overline{14} + \overline{42}$ are completed to 1 by $\overline{3} + \overline{14}$, since $\overline{3} + \overline{7} + \overline{42} = \overline{2}$. Multiplying $\overline{3} + \overline{14}$ by $\overline{67}$ gives the following alternative expression:

$$\overline{67} \times 2 = \overline{42} + \overline{201} + \overline{938}$$

The RMP value is superior on account of the smaller fractional number in its last term.

Partitioning into whole numbers in the manner described above will yield $2\overline{n}$ values for any value of n apart from $n = 5$. It is possible to write out the procedure in general terms as follows:

1) Choose a number greater than half n that has several factors (the 'chosen number').
2) Obtain the difference between this number and n.
3) Partition this difference into components that are factors of the chosen number.
4) Divide these components by the chosen number to obtain a series of fractions.
5) Subtract this series from 1 to obtain not more than three unit fractions.
6) Divide these fractions by n.
7) Precede them by the reciprocal of the chosen number.

Using $n = 67$ again as an example:

1) 40
2) $67 - 40 = 27$
3) $27 = 20 + 5 + 2$
4) $\overline{40}(20 + 5 + 2) = \overline{2} + \overline{8} + \overline{20}$
5) $1 - (\overline{2} + \overline{8} + \overline{20}) = \overline{5} + \overline{8}$
6) $\overline{67}(\overline{5} + \overline{8}) = \overline{335} + \overline{536}$
7) $\overline{67} \times 2 = \overline{40} + \overline{335} + \overline{536}$

It is plausible that the above procedure is the one used by the scribe, since it yields just the information that is supplied by the RMP, which for $n = 67$, for instance, simply states that $67 \times \overline{40} = 1 + \overline{2} + \overline{8} + \overline{20}$, $67 \times \overline{335} = \overline{5}$ and $67 \times \overline{536} = \overline{8}$.

The task of finding a suitable first fraction for the partition method and subsequent fractions that will complete to 1 is in practice less formidable than it sounds. Below is a list of first fractions, eight in all, related to specific values of n:

$\overline{20}$	$n = 31$
$\overline{30}$	$n = 35, 47, 53, 55$
$\overline{36}$	$n = 59$
$\overline{40}$	$n = 61, 67, 71$
$\overline{42}$	$n = 43$
$\overline{56}$	$n = 97$
$\overline{60}$	$n = 73, 79, 83, 89$
$\overline{70}$	$n = 91$

The ten fractional series summing to 1 that are needed to obtain the RMP expressions for $2\overline{n}$ when n has the above values are as follows:

$\overline{3} + \overline{5} + \overline{10} + \overline{30}$	$n = 91$
$\overline{3} + \overline{6} + \overline{6}$	$n = 35, 55$
$\overline{3} + \overline{6} + \overline{10} + \overline{15}$	$n = 53$
$\overline{2} + \overline{3} + \overline{7} + \overline{42}$	$n = 43$
$\overline{2} + \overline{3} + \overline{10} + \overline{15}$	$n = 47, 79$

$\overline{2} + \overline{4} + \ \overline{5} + \overline{20}$ $n = 31, 67$
$\overline{2} + \overline{4} + \ \overline{7} + \overline{14} + \overline{28}$ $n = 97$
$\overline{2} + \overline{4} + \ \overline{8} + \overline{10} + \overline{40}$ $n = 61, 71$
$\overline{2} + \overline{4} + \ \overline{9} + \overline{12} + \overline{18}$ $n = 59$
$\overline{3} + \overline{4} + \ \overline{5} + \ \overline{6} + \overline{20}$ $n = 73, 83, 89$

The partition method can be used to find values for $2\overline{n}$ beyond those given in the RMP — that is, when n is greater than 101. The case of $\overline{101} \times 2$ gives a special application of the partition method in which the chosen number in step 1 is equal to n and step 5 yields $1 = \overline{2} + \overline{3} + \overline{6}$. As already mentioned, there is no alternative to the RMP value if $\overline{101} \times 2$ is to have not more than four fractional numbers all less than 1000. Here is one value obtained by the partition method where this condition has been waived:

1) 60
2) $101 - 60 = 41$
3) $41 = 30 + 6 + 5$
4) $\overline{60}(30 + 6 + 5) = \overline{2} + \overline{10} + \overline{12}$
5) $1 - (\overline{2} + \overline{10} + \overline{12}) = \overline{4} + \overline{15}$
6) $\overline{101}(\overline{4} + \overline{15}) = \overline{404} + \overline{1515}$
7) $\overline{101} \times 2 = \overline{60} + \overline{404} + \overline{1515}$

A just acceptable value for $\overline{103} \times 2$ can be got by partition as follows:

1) 72
2) $103 - 72 = 31$
3) $31 = 24 + 4 + 3$
4) $\overline{72}(24 + 4 + 3) = \overline{3} + \overline{18} + \overline{24}$
5) $1 - (\overline{3} + \overline{18} + \overline{24}) = \overline{3} + \overline{8} + \overline{9}$
6) $\overline{103}(\overline{3} + \overline{8} + \overline{9}) = \overline{309} + \overline{824} + \overline{927}$
7) $\overline{103} \times 2 = \overline{72} + \overline{309} + \overline{824} + \overline{927}$

Other methods
There is another partition method for obtaining $2\overline{n}$. In essence it was propounded first by Hultsch in 1895, and was elaborated in 1952 and 1957 by Bruins, who was apparently unaware of Hultsch's earlier work. Bruins himself claimed inspiration from reading Plato's *Laws*, a passage from which is quoted on the title page of this book. A slightly modified formulation of the method is as follows:

1) Choose a number that is greater than half n and has several factors.
2) Obtain the difference between this 'chosen number' and n.
3) Subtract the difference from the chosen number.
4) Partition this result into not more than three components that are factors of the chosen number.
5) Divide these components by the chosen number.
6) Divide the fractions so obtained by n.
7) Precede them by the reciprocal of the chosen number.

It will be seen that steps 1, 2, 6 and 7 are the same as in the procedure outlined in the previous section. The essential difference is that the partitioning is done at a later stage, at step 4 rather than step 3. Taking $\overline{67} \times 2$ once more as an example, the stages go:

1) 40
2) $67 - 40 = 27$
3) $40 - 27 = 13$
4) $13 = 8 + 5$
5) $\overline{40}(8 + 5) = \overline{5} + \overline{8}$
6) $\overline{67}(\overline{5} + \overline{8}) = \overline{335} + \overline{536}$
7) $\overline{67} \times 2 = \overline{40} + \overline{335} + \overline{536}$

The operations as outlined above are extremely quick and easy to manage, since they escape the necessity of completing a fractional series to 1. But, by the same token, they fail to give information contained in the RMP that was clearly vital to the Egyptian method of calculating $2\overline{n}$. In the case of $n = 67$, the scribe explicitly states that a fortieth part of 67 is $1 + \overline{2} + \overline{8} + \overline{20}$, but the Hultsch–Bruins method does not involve this information.

A further small piece of evidence may be put forward suggesting that the Hultsch–Bruins procedure was not the one used by the ancient Egyptians. It arises in the calculation for $n = 91$. As we have formulated it, the various stages become:

1) 70
2) $91 - 70 = 21$
3) $70 - 21 = 49$
4) $49 = 35 + 14$
5) $\overline{70}(35 + 14) = \overline{2} + \overline{5}$
6) $\overline{91}(\overline{2} + \overline{5}) = \overline{182} + \overline{455}$
7) $\overline{91} \times 2 = \overline{70} + \overline{182} + \overline{455}$

In order to obtain the RMP expression for $\overline{91} \times 2$, it is necessary to factorise 91 into 13×7 and to equate $\overline{2} + \overline{5}$ to $\overline{10} \times 7$, giving

6) $(\overline{13} \times \overline{7})(\overline{10} \times 7) = \overline{130}$
7) $\overline{91} \times 2 = \overline{70} + \overline{130}$

Now, the expression $\overline{2} + \overline{5}$, which is critical to the Hultsch–Bruins calculation, does not occur in the RMP text. Instead the scribe used the alternative $\overline{3} + \overline{30}$, also equal to $\overline{10} \times 7$ but, one would have thought, rather less attractive than $\overline{2} + \overline{5}$ since it involves a larger fractional number. The fractions $\overline{3} + \overline{30}$ cannot be got directly from the partition in step 5. It can, however, arise naturally as part of a series summing to 1, namely $\overline{3} + \overline{5} + \overline{10} + \overline{30}$. The evaluation of $\overline{91} \times 2$ by the partition method that we think likely to have been used — at least for high values of n — goes as follows:

1) 70
2) $91 - 70 = 21$
3) $21 = 14 + 7$
4) $\overline{70}(14 + 7) = \overline{5} + \overline{10}$
5) $1 - (\overline{5} + \overline{10}) = \overline{3} + \overline{30}$
6) $\overline{91}(\overline{3} + \overline{30}) = (\overline{13} \times \overline{7})(\overline{10} \times 7) = \overline{130}$
7) $\overline{91} \times 2 = \overline{70} + \overline{130}$

The basis of the Hultsch–Bruins method is that the chosen number is effectively doubled, thereby generating the 2 of $2\bar{n}$ without completing a fractional series to 1. In the version described above, the doubling occurs at step 3. For moderns, less conversant with series summing to 1 than were the ancient Egyptians, the method provides the simplest way of arriving at alternative expressions for $2\bar{n}$. Gillings, followed by Bruckheimer and Salomon, ran computer programs aiming to obtain a complete set of such expressions; many of these could, however, have been got quite simply using either partition method.

Yet another method has been suggested for the determination of $2\bar{n}$, by Gillings in 1974. He manipulated the terms of the identity $2\bar{n} = \bar{n}(1 + \bar{2} + \bar{3} + \bar{6})$ to obtain the RMP value of $2\bar{n}$ for any value of n. In the case of $n = 67$ the steps were as follows:

$$
\begin{aligned}
\bar{2} + \bar{3} + \bar{6} &= \bar{4} + \bar{4} + \bar{2}\\
&= (\bar{5} + \overline{20}) + (\bar{8} + \bar{8}) + \bar{2}\\
&= (\bar{2} + \bar{8} + \overline{20}) + (\bar{5} + \bar{8})\\
2 = 1 + \bar{2} + \bar{3} + \bar{6} &= (1 + \bar{2} + \bar{8} + \overline{20}) + (\bar{5} + \bar{8})\\
\overline{67} \times 2 &= \overline{67}(1 + \bar{2} + \bar{8} + \overline{20}) + \overline{67}(\bar{5} + \bar{8})\\
&= \overline{40} + \overline{335} + \overline{536}
\end{aligned}
$$

It still has to be proved that $\overline{67}(1 + \bar{2} + \bar{8} + \overline{20}) = \overline{40}$, which, as we have shown, is most readily done by the partition method. It is not clear that the manipulation of $\bar{2} + \bar{3} + \bar{6}$ is anything more than an elaborate way of producing, with hindsight, a suitable series that sums to 1. Furthermore, the various stages bear no relation to such working as is included in the RMP.

Conclusions and table

In considering what methods the Egyptians may have used to double unit fractions, it must be emphasised again that it is essential to take note of such working as is included in the text, and to be cautious about erecting any theory that does not have some attestation. On the other hand, it must be borne in mind that RMP entries may serve different purposes. They may indicate the actual methods employed by the scribe, they may offer formal proofs of what he has previously worked out by other means, or they may represent the procedures that an apprentice was expected to go through as part of his scribal training so as to engender in him a better understanding of mathematical principles and practice.

Above $n = 29$, except when $n = 37$ and 41, no detailed working is given. Undoubtedly, for higher values of n, actual divisions would have become increasingly tedious to record in full as well as to perform, and it seems likely that by the time $n = 31$ was reached the pattern of the results obtained up to that point would have suggested partitioning n as a short cut. This would have produced, in an expeditious way, a means of obtaining alternative values for $2\bar{n}$ that could, if necessary, be assessed on their merits. Since the partition method is applicable to any value of n (not strictly $n = 5$, since it does not split into whole numbers: see the ensuing table), it would be possible, if desired, to go back to the smaller values of n and use it to check that the results achieved by division were indeed the most appropriate.

The criteria for selection were usually, though not always, that the smallest possible number of fractions should appear in the expression for $2\bar{n}$, and that fractional numbers that were odd or needlessly large should be avoided. The RMP choice seems to have remained for the most part canonical for a long time. However, a variant for $\bar{7} \times 2$ has turned up on an 18th-dynasty ostracon (no. 153 from Theban tomb no. 71) with the RMP value $\overline{4} + \overline{28}$ replaced by $\overline{6} + \overline{14} + \overline{21}$. The alternative values can be got as follows:

$\dfrac{1}{2}$	7	$\dfrac{1}{3}$	7
$\overline{/4}$	$3 + \overline{2}$	$\overline{3}$	$4 + \overline{\overline{3}}$
	$1 + \overline{2} + \overline{4}$	$\overline{3}$	$2 + \overline{3}$
		$\overline{/6}$	$1 + \overline{6}$
Remainder	$\overline{4}$	Remainder	$\overline{2} + \overline{3}$
2	14	$\overline{/2}$	14 $(\overline{14})$
$\overline{/4}$	28 $(\overline{28})$	$\overline{/3}$	21 $(\overline{21})$

giving

$$\overline{7} \times 2 = \overline{4} + \overline{28}$$

giving

$$\overline{7} \times 2 = \overline{6} + \overline{14} + \overline{21}$$

It appears that whoever devised this variant for $\overline{7} \times 2$ was trying the effect of using terms of the $\overline{\overline{3}}, \overline{3}, \overline{6}$ series instead of the $\overline{2}, \overline{4}, \overline{8}$ series as multipliers. The RMP version can be said to be superior on two counts: there are only two terms instead of three, and both of them have even numbers.

Although, on the criteria mentioned above, it is usually clear which of several alternative values for $2\overline{n}$ is to be preferred, there are times when relative merits conflict: $\overline{31} \times 2$ and $\overline{43} \times 2$, already considered, are cases in point. In such instances it is possible that the scribe might have been influenced in his choice by the ease or elegance of the partitioning of n. The stages of the calculation, according to our method of setting them out, when $n = 31$ and $n = 43$, are as follows:

1) 20	1) 42
2) $31 - 20 = 11$	2) $43 - 42 = 1$
3) $11 = 10 + 1$	3) —
4) $\overline{20}(10 + 1) = \overline{2} + \overline{20}$	4) $\overline{42} \times 1 = \overline{42}$
5) $1 - (\overline{2} + \overline{20}) = \overline{4} + \overline{5}$	5) $1 - \overline{42} = \overline{2} + \overline{3} + \overline{7}$
6) $\overline{31}(\overline{4} + \overline{5}) = \overline{124} + \overline{155}$	6) $\overline{43}(\overline{2} + \overline{3} + \overline{7}) = \overline{86} + \overline{129} + \overline{301}$
7) $\overline{31} \times 2 = \overline{20} + \overline{124} + \overline{155}$	7) $\overline{43} \times 2 = \overline{42} + \overline{86} + \overline{129} + \overline{301}$

The decomposition of 31 into $20 + 10 + 1$ and of 43 into $42 + 1$ is neat and may have proved compellingly satisfying.

The following table shows the partitions producing RMP values of $2\overline{n}$ for all values of n from 5 to 99. The relative frequency with which the first fractional number is a multiple of 10 can be explained by the fact that it is associated with a partitioning of n into tens and units, resulting naturally from the Egyptian decimal system.

n	$2\overline{n}$	n	$2\overline{n}$
$5 = 3 + (1 + \overline{2}) + \overline{2}$	$\overline{3} + (\overline{5} \times \overline{3})$	$53 = 30 + 20 + 3$	$\overline{30} + \overline{53}(\overline{6} + \overline{15})$
$7 = 4 + 2 + 1$	$\overline{4} + (\overline{7} \times \overline{4})$	$55 = 30 + 20 + 5$	$\overline{30} + (\overline{55} \times \overline{6})$
$9 = 6 + 3$	$\overline{6} + (\overline{9} \times \overline{2})$	$57 = 38 + 19$	$\overline{38} + (\overline{57} \times \overline{2})$
$11 = 6 + 4 + 1$	$\overline{6} + (\overline{11} \times \overline{6})$	$59 = 36 + 18 + 3 + 2$	$\overline{36} + \overline{59}(\overline{4} + \overline{9})$
$13 = 8 + 4 + 1$	$\overline{8} + \overline{13}(\overline{4} + \overline{8})$	$61 = 40 + 20 + 1$	$\overline{40} + \overline{61}(\overline{4} + \overline{8} + \overline{10})$

$15 = 10+5$	$\overline{10}+(\overline{15}\times\overline{2})$	$63 = 42+21$	$\overline{42}+(\overline{63}\times\overline{2})$
$17 = 12+4+1$	$\overline{12}+\overline{17}(\overline{3}+\overline{4})$	$65 = 39+26$	$\overline{39}+(\overline{65}\times\overline{3})$
$19 = 12+6+1$	$\overline{12}+\overline{19}(\overline{4}+\overline{6})$	$67 = 40+20+5+2$	$\overline{40}+\overline{67}(\overline{5}+\overline{8})$
$21 = 14+7$	$\overline{14}+(\overline{21}\times\overline{2})$	$69 = 46+23$	$\overline{46}+(\overline{69}\times\overline{2})$
$23 = 12+8+3$	$\overline{12}+(\overline{23}\times\overline{12})$	$71 = 40+20+10+1$	$\overline{40}+\overline{71}(\overline{8}+\overline{10})$
$25 = 15+10$	$\overline{15}+(\overline{25}\times\overline{3})$	$73 = 60+10+3$	$\overline{60}+\overline{73}(\overline{3}+\overline{4}+\overline{5})$
$27 = 18+9$	$\overline{18}+(\overline{27}\times\overline{2})$	$75 = 50+25$	$\overline{50}+(\overline{75}\times\overline{2})$
$29 = 24+4+1$	$\overline{24}+\overline{29}(\overline{2}+\overline{6}+\overline{8})$	$77 = 44+22+11$	$\overline{44}+(\overline{77}\times\overline{4})$
$31 = 20+10+1$	$\overline{20}+\overline{31}(\overline{4}+\overline{5})$	$79 = 60+15+4$	$\overline{60}+\overline{79}(\overline{3}+\overline{4}+\overline{10})$
$33 = 22+11$	$\overline{22}+(\overline{33}\times\overline{2})$	$81 = 54+27$	$\overline{54}+(\overline{81}\times\overline{2})$
$35 = 30+5$	$\overline{30}+\overline{35}(\overline{3}+\overline{6})$	$83 = 60+20+3$	$\overline{60}+\overline{83}(\overline{4}+\overline{5}+\overline{6})$
$37 = 24+12+1$	$\overline{24}+\overline{37}(\overline{3}+\overline{8})$	$85 = 51+34$	$\overline{51}+(\overline{85}\times\overline{3})$
$39 = 26+13$	$\overline{26}+(\overline{39}\times\overline{2})$	$87 = 58+29$	$\overline{58}+(\overline{87}\times\overline{2})$
$41 = 24+16+1$	$\overline{24}+\overline{41}(\overline{6}+\overline{8})$	$89 = 60+20+6+3$	$\overline{60}+\overline{89}(\overline{4}+\overline{6}+\overline{10})$
$43 = 42+1$	$\overline{42}+\overline{43}(\overline{2}+\overline{3}+\overline{7})$	$91 = 70+14+7$	$\overline{70}+\overline{91}(\overline{3}+\overline{30})$
$45 = 30+15$	$\overline{30}+(\overline{45}\times\overline{2})$	$93 = 62+31$	$\overline{62}+(\overline{93}\times\overline{2})$
$47 = 30+15+2$	$\overline{30}+\overline{47}(\overline{3}+\overline{10})$	$95 = 60+30+5$	$\overline{60}+\overline{95}(\overline{4}+\overline{6})$
$49 = 28+14+7$	$\overline{28}+(\overline{49}\times\overline{4})$	$97 = 56+28+7+4+2$	$\overline{56}+\overline{97}(\overline{7}+\overline{8})$
$51 = 34+17$	$\overline{34}+(\overline{51}\times\overline{2})$	$99 = 66+33$	$\overline{66}+(\overline{99}\times\overline{2})$

The pattern of the above table is as follows. Each expression for $2\overline{n}$ consists of a first fraction which is the reciprocal of the first term in the decomposition of n, together with the product of $\overline{n}$ and one or more fractions which, when added to the product of the first fraction and the components of n, form a series that sums to 2.

The first six numbered items in the RMP, which occur on the recto of the New York fragments and BM 10057 immediately after the table for doubling fractions, are concerned with the equal distribution among ten men of different numbers of loaves from 1 to 9, excluding, for some reason, 3, 4 and 5. The answers in each case consist of one, two or three fractions, the smallest being $\overline{30}$. It is difficult to imagine loaves being divided into thirtieths as a practical procedure, but there are, in fact, documentary records of fractions of loaves being issued to temple employees on the basis of so many loaves being available for a certain number of men. Apart from furnishing simple examples of the proper allocation when payment was made in kind, as it always was, RMP nos. 1–6 would also have been useful in computation generally, since 10 often occurs as a divisor. The results that would have been particularly helpful are those where the dividend is not a factor of 10:

$$6 \div 10 = \bar{2} + \overline{10}$$
$$7 \div 10 = \bar{3} + \overline{30}$$
$$8 \div 10 = \bar{3} + \overline{10} + \overline{30}$$
$$9 \div 10 = \bar{3} + \bar{5} + \overline{30}$$

Fig.6. *Baking bread, from the mastaba of Khentika called Ikhekhi at Saqqara.*

No working is included in the RMP to show how any of the above identities were obtained. On the other hand, in all cases, including the examples where the dividends are 1 and 2, proofs working back from the answer to the problem are given in full. They are probably intended as exercises in the multiplication of fractions to be carried out by successive doubling. The actual answers could easily have been got by fractional multiplication aided by partition of the dividend as follows:

1 6 = 5+1	1 7 = 5+2	1 8 = 5+2+1	1 9 = 5+2+2
$\bar{5}$ 1+$\bar{5}$	$\bar{5}$ 1+($\bar{3}$+$\overline{15}$)	$\bar{5}$ 1+($\bar{3}$+$\overline{15}$)+$\bar{5}$	$\bar{5}$ 1+($\bar{3}$+$\overline{15}$)+($\bar{3}$+$\overline{15}$)
	= (1+$\bar{3}$)+$\overline{15}$	= (1+$\bar{3}$)+$\bar{5}$+$\overline{15}$	= (1+$\bar{3}$)+($\bar{3}$+$\overline{15}$)+$\overline{15}$
$\overline{10}$ 2+$\overline{10}$	$\overline{10}$ 3+$\overline{30}$	$\overline{10}$ 3+$\overline{10}$+$\overline{30}$	$\overline{10}$ 3+$\bar{5}$+$\overline{30}$

The identity $\bar{5} \times 2 = \bar{3} + \overline{15}$, used several times above, is taken from the table of doubled fractions.

The problems immediately following, nos. 7–20 and 21–23, dealing respectively with the use of common multipliers and with completion of fractional series, have already been considered in the section on addition of fractions (page 19).

SOLUTION OF EQUATIONS AND RELATED TABLES: RMP nos. 24–38, 47, 80–81

These problems are numerical exercises involving an unknown, and so represent an approach towards algebra. Although algebraic symbols were not used, the unknown quantity being designated as such verbally, it is convenient for the purposes of exposition to adopt the modern practice of calling the unknown 'x'. Nos. 24–27 are more easily solved since the coefficient of x has only two terms, whereas in nos. 30–34 it has three or four. Nos. 28 and 29 will be considered in the later section 'Diversions' (page 54). The simpler equations are:

no. 24	$x + \overline{7}x = 19$
no. 25	$x + \overline{2}x = 16$
no. 26	$x + \overline{4}x = 15$
no. 27	$x + \overline{5}x = 21$

It is required in these equations to add to the unknown quantity a fraction of itself so that the sum is equal to a given whole number. The procedure is to eliminate the fraction by putting the fractional number in place of the unknown. Multipliers are then found for the result so that the partial products sum to the given number. Finally, these multipliers are themselves multiplied by the fractional number to obtain the solution. In the case of no. 24, the stages of the calculation are as follows:

1) Insertion of 7 in the place of the unknown yields 8:
   $7 + (7 \times \overline{7}) = 8$

2) It is found that 8 has to be multiplied by $2 + \overline{4} + \overline{8}$ to get 19:
   $8(2 + \overline{4} + \overline{8}) = 16 + 2 + 1 = 19$

3) 7 has to be increased by a factor of $(2 + \overline{4} + \overline{8})$ to give 19 when substituted for the unknown:
   $7(2 + \overline{4} + \overline{8}) = 16 + \overline{2} + \overline{8}$ (Answer.)

The procedure outlined above is best seen as an exercise in proportion following the removal of the fraction. Proportion is a concept that was certainly known to the Egyptians. The method of solving has been seen as an example of 'false supposition', in which a 'trial number' is inserted into the equation and afterwards adjusted. It must be realised, however, that the choice of 'trial number' is not arbitrary but is dictated by the fraction. In step 3, the 'trial number' should, logically, be multiplied by the proportional factor to get the answer. In the RMP it is the other way round: the proportional factor is multiplied by the 'trial number', for the good reason that the latter is a whole number, and so more convenient to operate with than a fraction. For instance, it is conceptually easier to grasp that $\bar{2} + \bar{4} + \bar{8}$ is equal to seven times an eighth than to an eighth of seven, since the former can be got directly by doubling and adding but the latter involves a division. In matters such as this, the scribe was less interested in logic than in having an effective method that worked.

The harder equations to be solved are:

no. 31 $x + (\bar{\bar{3}} + \bar{2} + \bar{7})x = 33$
no. 32 $x + (\bar{\bar{3}} + \bar{4})x \quad\quad = 2$
no. 33 $x + (\bar{3} + \bar{2} + \bar{7})x = 37$
no. 34 $x + (\bar{2} + \bar{4})x \quad\quad = 10$

Nos. 31 and 33 are similar in that they require the same coefficient of x to be the divisor in a division sum. Of the two, no. 31 turns out to be slightly more awkward; indeed, the numbers seem to have been chosen with the purpose of making the sum difficult. It is required to find a quantity that, when added to two-thirds, a half and a seventh of itself, becomes 33; in other words, to solve $(1 + \bar{\bar{3}} + \bar{2} + \bar{7})x = 33$. If confronted with the equation today, one would express the coefficient of x as the fraction $97/42$ and calculate $33 \times 42/97$. The answer, expressed as a decimal correct to two places, is 14.29, slightly in excess of $14\frac{1}{4}$.

The Egyptian scribe came to a similar conclusion. He successively doubled $1 + \bar{\bar{3}} + \bar{2} + \bar{7}$ until he had multipliers summing to 14. He then found that a further multiplication by one-quarter brought the partial products to a total just below 33. In order to simplify the task of adding a long string of fractions, he took the smaller ones separately. The calculation then proceeded as follows:

$$
\begin{aligned}
(1 + \bar{\bar{3}} + \bar{2} + \bar{7})(2 + 4 + 8 + \bar{4}) &= 32 + \bar{2} + (\bar{7} + \bar{8} + \overline{14} + \overline{28} + \overline{28}) \\
&= 32 + \bar{2} + \overline{42}(17 + \bar{4}) \\
&= 32 + \bar{2} + \overline{42}[21 - (3 + \bar{2} + \bar{4})] \\
&= 33 - \overline{42}(3 + \bar{2} + \bar{4}).
\end{aligned}
$$

The next step is to express $\overline{42}(3 + \bar{2} + \bar{4})$ as a multiple of $1 + \bar{\bar{3}} + \bar{2} + \bar{7}$. This is achieved through equating $1 + \bar{\bar{3}} + \bar{2} + \bar{7}$ to $\overline{42} \times 97$, so that $\overline{42} = (1 + \bar{\bar{3}} + \bar{2} + \bar{7}) \times \overline{97}$ and the required multiple is $(1 + \bar{\bar{3}} + \bar{2} + \bar{7}) \times \overline{97}(3 + \bar{2} + \bar{4})$. Then, since $2 + 4 + 8 + \bar{4} = 14 + \bar{4}$, the previous calculation, given above, is transformed by the appropriate substitutions into

$$(1 + \bar{\bar{3}} + \bar{2} + \bar{7})\,[14 + \bar{4} + \overline{97}(3 + \bar{2} + \bar{4})] = 33$$

Since $\overline{97} \times 2 = \overline{56} + \overline{679} + \overline{776}$ from the table for doubling fractions, the answer becomes

$$14 + \bar{4} + \overline{56} + \overline{97} + \overline{194} + \overline{338} + \overline{679} + \overline{776}$$

The answer to RMP no. 33 is obtained in a similar fashion. It is required to solve $(1 + \bar{\bar{3}} + \bar{2} + \bar{7})x = 37$. First $(1 + \bar{\bar{3}} + \bar{2} + \bar{7}) \times 16$ is found to exceed 37 by $\overline{42} \times 2$. Now $\overline{42} \times 2$ expressed as a multiple of $1 + \bar{\bar{3}} + \bar{2} + \bar{7}$ is equal to $(1 + \bar{\bar{3}} + \bar{2} + \bar{7}) \times \overline{97} \times 2$. It follows that

$$(1 + \bar{\bar{3}} + \bar{2} + \bar{7})(16 + \bar{56} + \bar{679} + \bar{776}) = 37$$

It will be noted that the technique of solving the equation of RMP nos. 31 and 33 is analogous to the basic method used for doubling fractions. The calculation involves finding a dividend that, with the coefficient of x rather than n as divisor, is close to the whole number, which now is 33 or 37 instead of 2. The difference is divided by the coefficient of x to complete the answer.

The treatment of RMP no. 32 is similar, except that the multipliers in the first stage belong to the $\bar{\bar{3}}, \bar{3}, \bar{6}$ series rather than the doubling/halving series. It is required to find a quantity that, when added to one-third and a quarter of itself, becomes 2, in other words, to solve $(1 + \bar{3} + \bar{4})x = 2$. The first stage of the calculation gives

$$(1 + \bar{3} + \bar{4})(\bar{\bar{3}} + \bar{3} + \bar{6} + \bar{12}) = \bar{144}(288 - 3)$$
$$= 2 - (\bar{144} \times 3)$$

But $144(1 + \bar{3} + \bar{4}) = 144 + 48 + 36 = 228$, so that $\bar{144} = (1 + \bar{3} + \bar{4}) \times \bar{228}$ and $\bar{144} \times 3 = \bar{144}(2 + 1) = (1 + \bar{3} + \bar{4})(\bar{114} + \bar{228})$. It follows that

$$(1 + \bar{3} + \bar{4})(\bar{\bar{3}} + \bar{3} + \bar{6} + \bar{12} + \bar{114} + \bar{228}) = 2$$

giving the answer:

$$1 + \bar{6} + \bar{12} + \bar{114} + \bar{228}$$

The scribe was presumably more concerned with the method than with the result, and so did not bother to simplify it. Since $\bar{6} + \bar{12} = \bar{4}$ and $\bar{114} + \bar{228} = \bar{19}(\bar{6} + \bar{12}) = \bar{19} \times \bar{4} = \bar{76}$, his solution could have been reduced to:

$$(1 + \bar{3} + \bar{4})(1 + \bar{4} + \bar{76}) = 2$$

The simplified answer could have been got directly, by multiplying $1 + \bar{3} + \bar{4}$ not by terms of the $\bar{\bar{3}}, \bar{3}, \bar{6}$ series but by $1 + \bar{4}$.

In RMP no. 34, the numbers are such that an answer could be got directly by division. It is required to find a quantity that added to its half and its quarter becomes 10, in other words, to solve $(1 + \bar{2} + \bar{4})x = 10$. The calculation is as follows:

/1	$1 + \bar{2} + \bar{4}$
2	$3 + \bar{2}$
/4	7
/$\bar{7}$	$\bar{4}$
$\bar{4} + \bar{28}$	$\bar{2}$
/$\bar{2} + \bar{14}$	1

giving

$$(1 + \bar{2} + \bar{4})(5 + \bar{2} + \bar{7} + \bar{14}) = 10$$

The choice of multipliers, involving successive doubling and the introduction of a seventh, is an obvious one in view of the identity $1 + \bar{2} + \bar{4} = \bar{4} \times 7$.

The answer to no. 34 could have been got from an earlier RMP problem, no. 9, which gives the reciprocal of $1 + \bar{2} + \bar{4}$ as $\bar{2} + \overline{14}$. This reciprocal multiplied by 10 is equal to $5 + \bar{2} + \bar{7} + \overline{14}$.

Problems 35–38 involve first-degree equations as applied to subdivisions of a *hekat*. Ordinary fractions of a *hekat* had to be re-expressed in Horus-eye fractions and *ro*. It will be remembered that the Horus-eye fractions of a *hekat* are $\bar{2}, \bar{4}, \bar{8}, \overline{16}, \overline{32}, \overline{64}$ and that a *ro* is equal to one-tenth of the $\overline{32}$ fraction, and so to $\overline{320}$ of a *hekat*. It is evident from RMP no. 35 that the Horus-eye fractions were obtained by first converting *hekat* to *ro* and then back to Horus-eye fractions and *ro*. In this problem, it was required to solve $(3 + \bar{3})x = 1$ *hekat*. Division of 1 by $3 + \bar{3}$ yielded the answer $\bar{5} + \overline{10}$ *hekat*. Converted to *ro*, this is $320(\overline{10} + \bar{5}) = 32 + 64 = 96$ *ro*. A table such as the following could have been constructed to aid conversion from *ro* to Horus-eye fractions:

$5\ ro$	$= \overline{64}\ hekat$
$10\ ro$	$= \overline{32}\ hekat$
$20\ ro$	$= \overline{16}\ hekat$
$40\ ro$	$= \bar{8}\ hekat$
$80\ ro$	$= \bar{4}\ hekat$
$160\ ro$	$= \bar{2}\ hekat$

Fig.7. *Recording the amount of grain brought to the granary, from the tomb of Antefoker at Thebes.*

which gives $96\ ro = (80 + 10 + 5) + 1\ ro = \bar{4} + \overline{32} + \overline{64}\ hekat + 1\ ro$. This is yet another example of how a problem can be solved by partition.

Other conversion tables are provided in RMP nos. 47, 80 and 81. No. 47 gives one-tenth gradations of 100 quadruple *hekat* expressed in Horus-eye fractions of a quadruple *hekat* and quadruple *ro* as applied to grain in a circular or rectangular granary. The shape, of course, is irrelevant and is put in only because the table is an adjunct to a section of the RMP dealing with the volume of different types of granary. A curious feature of the table is that $\overline{70} \times 100$ *hekat* is translated into $1 + \bar{4} + \bar{8} + \overline{32} + \overline{64}$ *hekat* and $2 + \overline{14} + \overline{21} + \overline{42}$ *ro*. But $\overline{14} + \overline{21} + \overline{42} = \bar{7}(\bar{2} + \bar{3} + \bar{6}) = \bar{7}$, so the additional number of *ro* should have been given as $2 + \bar{7}$. The calculation should go as follows:

$$\overline{70} \times 100 = \bar{7} \times 10 = 1 + \bar{4} + (\bar{7} + \overline{28})\ hekat$$
$$\bar{7} + \overline{28}\ hekat = \bar{7} \times 400 = 57 + \bar{7}\ ro$$
$$= \bar{8} + \overline{32} + \overline{64}\ hekat + 2 + \bar{7}\ ro$$

so that

$$\overline{70} \times 100\ \text{hekat} = 1 + \bar{4} + \bar{8} + \overline{32} + \overline{64}\ hekat + 2 + \bar{7}\ ro$$

Problems nos. 80 and 81 relate Horus-eye fractions of the *hekat* to the *hin*, which is the liquid or grain measure equal to one-tenth of a *hekat*.

UNEQUAL DISTRIBUTION OF GOODS AND OTHER PROBLEMS
RMP nos. 39–40, 61, 63–65, 67–68

In RMP no. 39, 100 loaves are to be divided between 10 men, 50 being shared equally among 6, and 50 among 4. It is required to find the difference of the shares. The problem is solved in the obvious manner by two division sums: $50 \div 4 = 12 + \bar{2}$ and $50 \div 6 = 8 + \bar{3}$. The difference of share $= (12 + \bar{2}) - (8 + \bar{3}) = 4 + \bar{6}$.

In RMP no. 65, 100 loaves are distributed to 10 men, 3 of whom are to have double. To add interest, the information is provided that the three lucky ones were a sailor, a foreman and a watchman. The solution is as follows. The 100 loaves are divided into $10 + 3 = 13$ portions. The number of loaves in a single portion, given by $100 \div 13$, is $7 + \bar{3} + \overline{39}$. For details of this calculation, see our earlier section on 'multiplication and division' (page 16). The number of loaves in a double portion, therefore, is equal to $14 + (1 + \bar{3}) + (\overline{39} \times 2)$, or $15 + \bar{3} + \overline{26} + \overline{78}$.

In RMP no. 63, 700 loaves are to be divided among 4 men in the ratio $\bar{\bar{3}}:\bar{2}:\bar{3}:\bar{4}$ and it is asked how many loaves each man is to get. The procedure is to add the proportions, divide the number to be distributed by their sum, and multiply the result by each of the proportions in turn. This is the same as the modern method. Suppose the men to have $\bar{\bar{3}}k, \bar{2}k, \bar{3}k$ and $\bar{4}k$ loaves, where k is a common factor: then $k(\bar{\bar{3}} + \bar{2} + \bar{3} + \bar{4}) = 700$, so that $k = 700 \div (\bar{\bar{3}} + \bar{2} + \bar{3} + \bar{4}) = 700 \div (1 + \bar{2} + \bar{4})$. The scribe's working is as follows:

$$\bar{\bar{3}} + \bar{2} + \bar{3} + \bar{4} = 1 + \bar{2} + \bar{4}$$
$$1 \div (1 + \bar{2} + \bar{4}) = \bar{2} + \overline{14}$$
$$700(\bar{2} + \overline{14}) = 350 + 50 = 400$$

so that the answer in numbers of loaves is:

$$\bar{\bar{3}} \times 400 = 266 + \bar{\bar{3}}$$
$$\bar{2} \times 400 = 200$$
$$\bar{3} \times 400 = 133 + \bar{3}$$
$$\bar{4} \times 400 = 100$$

total   700

RMP no. 67 is not concerned with distribution but, like no. 63, involves the finding of a reciprocal. The problem is a rather artificial one relating to the numbering of cattle. The herdsman is accused of underpayment of tax since two-thirds of one-third of his herd was due. He protests that the required number is there, since he has brought 70.

The calculation shows that the number of the herd must then be 315, this being the solution of the equation $(\overline{3} \times \overline{\overline{3}})x = 70$. Following the rule enunciated in RMP no. 61B that multiplication of an odd-numbered fraction by $\overline{\overline{3}}$ is effected by equating it to $\overline{2} + \overline{6}$, $\overline{3} \times \overline{\overline{3}}$ can be replaced by $\overline{6} + \overline{18}$. The same result can, of course, be got from putting $\overline{3} \times \overline{\overline{3}}$ equal to $\overline{9} \times 2$. The reciprocal of $\overline{6} + \overline{18}$ is found by division to be $4 + \overline{2}$ as follows:

$$
\begin{array}{ll}
1 & \overline{6} + \overline{18} \\
2 & \overline{3} + \overline{9} \\
/4 & \overline{\overline{3}} + \overline{6} + \overline{18} \\
/2 & \overline{9}
\end{array}
$$

Total $4 + \overline{2}$ 1

since

$$\overline{6} + \overline{9} + \overline{18} = \overline{6} + \overline{3}(\overline{3} + \overline{6}) = \overline{6} + (\overline{3} \times \overline{2}) = \overline{3}$$

An easier method of reaching the same result would have been to proceed from $\overline{3} \times \overline{\overline{3}} = \overline{9} \times 2$. The reciprocal of this is $\overline{2} \times 9$, which is equal to $4 + \overline{2}$. The answer to the problem is got by multiplying 70 by $4 + \overline{2}$ to obtain 315.

In RMP no. 68, 4 overseers have received 100 quadruple *hekat* of grain. Their gangs consist of 12, 8, 6 and 4 men. It is required to discover how much each overseer gets to distribute. To solve this problem, 100 has to be divided into 4 parts in the ratio 12:8:6:4. The total number of men is equal to $12 + 8 + 6 + 4 = 30$, so that the amount available for each man is $100 \div 30 = 3 + \overline{3}$ quadruple *hekat*. The first overseer must receive $(3 + \overline{3}) \times 12 = 40$ quadruple *hekat*, the second overseer $(3 + \overline{3}) \times 8 = 26 + \overline{\overline{3}}$ quadruple *hekat*, the third overseer $(3 + \overline{3}) \times 6 = 20$ quadruple *hekat*, and the fourth overseer $(3 + \overline{3}) \times 4 = 13 + \overline{3}$ quadruple *hekat*.

The allocations to the second and fourth overseers were translated into Horus-eye fractions. This could have been done in the manner indicated for RMP no. 35 in the previous section. However, the RMP scribe, no doubt with the aim of giving his pupils the maximum practice in handling Horus-eye fractions took the longer route of converting the $3 + \overline{3}$ quadruple *hekat* that each man received into Horus-eye fractions and *ro* and then multiplying successively by 2, 4, and 8, summing the appropriate partial products to get $12\times$, $8\times$, $6\times$ and $4\times$. A resourceful pupil might have saved himself some labour by referring to the table in RMP no. 47, since it gives a value for $100 \div 30$ quadruple *hekat*, the amount for each man, in terms of Horus-eye fractions and *ro*.

RMP nos. 40 and 64 are of interest because they show that the Egyptians had an understanding of arithmetical progressions. In no. 40, 100 loaves are to be divided among 5 men so that the sum of the two smallest shares is one-seventh of the sum of the three greatest. It is evident from the working, although not stated, that the shares are to be in arithmetical progression. It is required to find the common difference of the shares.

The scribal solution to this problem made use of the descending arithmetical series of five terms that has 1 as its last term and sums, not to 100, but to 60. The middle term of this series is equal to the sum divided by the number of terms, and is, therefore, 12. The common difference is found by subtracting the last term from the middle term and halving the result, and so is equal to $(12 - 1) \times \bar{2} = 5 + \bar{2}$. Successively adding and subtracting the common difference to the middle term gives the complete series:

$$23 + (17 + \bar{2}) + 12 + (6 + \bar{2}) + 1 = 60$$

The sum of the three largest terms of the series is equal to three times the middle term plus three times the common difference, that is, $(12 \times 3) + [(5 + \bar{2}) \times 3] = 52 + \bar{2}$. The sum of the two smallest terms is equal to $7 + \bar{2}$. But $7 + \bar{2}$ is one-seventh of $52 + \bar{2}$, so that the series in this respect satisfies the required conditions. A series that satisfies the conditions completely, in that it sums to 100 instead of 60, can be got by proportion. The proportional factor is determined by finding out how many times 100 is greater than 60, in other words, by dividing 60 into 100:

$$/1 \qquad\qquad 60$$
$$/\bar{\bar{3}} \qquad\qquad 40$$

Total $1 + \bar{\bar{3}}$ 100

The common difference of the new series can be obtained by multiplying $5 + \bar{2}$ by $1 + \bar{\bar{3}}$ to get $9 + \bar{6}$. The complete series becomes:

$$(38 + \bar{3}) + (29 + \bar{6}) + 20 + (10 + \bar{\bar{3}} + \bar{6}) + (1 + \bar{\bar{3}}) = 100$$

In RMP no. 64 the problem is to distribute 10 *hekat* of barley among 10 men so that the shares are in arithmetical progression with a common difference of $\bar{8}$ *hekat*, and to discover how much each man gets. The scribe knew the rule that, to find the largest term of an arithmetical progression, he must add half the common difference to the average of the terms as many times as there are common differences, that is, one less than the number of terms. The mean share was 1 *hekat* (the copyist wrote $\bar{2}$ in error but here as elsewhere we have ignored trivial errors where the intention was obvious). Half the common difference was $\overline{16}$ *hekat*. The number of common differences was 9. The largest share, therefore, was $1 + (\overline{16} \times 9) = 1 + \overline{16}(8 + 1) = 1 + \bar{2} + \overline{16}$ *hekat*. The other shares are got by successively subtracting the common difference of $\bar{8}$ *hekat* from the largest share.

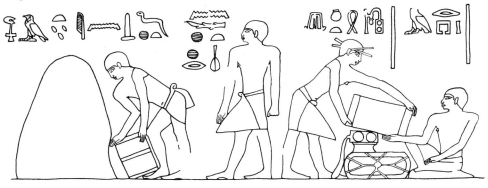

Fig.8. *Measuring barley and handing the accounts by the scribe of the granary to the steward, from the mastaba of Sekhemankhptah originally at Saqqara, now in the Museum of Fine Arts, Boston.*

SQUARING THE CIRCLE
RMP nos. 41–43, 48, 50

The Egyptians needed to be able to determine the capacity of cylindrical granaries. They knew that the volume of a cylinder was, like that of a rectangular container, equal to the area of its base multiplied by the height, so they had to be able to determine the area of the circular base. Their method for obtaining the area of a circle, as recorded in RMP nos. 41–43, was in fact the best in the prehellenic world. The instructions were simple: take the diameter of the circle, subtract its ninth part, and square the result to get the area.

The accuracy of the Egyptian procedure can be assessed by comparing the value obtained with the true value, which is equal to a quarter of the square of the diameter multiplied by the irrational number π, defined as the ratio of the circumference of a circle to its diameter. In RMP no. 41, the circular base of the granary whose volume is to be found has a diameter of 9 cubits. The area of its base according to the Egyptian rule is $[9 - (9 \times \overline{9})]^2 = 8^2 = 64$ square cubits, and since the height of the granary is 10 cubits, its volume is 640 cubic cubits. This volume is the same as $640(1 + \overline{2}) = 960\,khar$ or $960 \times \overline{20} = 48$ hundreds of quadruple *hekat* (92.6 cu m). The true volume is equal to $\pi \times$ one-quarter the square of the diameter multiplied by the height, which comes to 636 cubic cubits to the nearest whole number. The error is 4 cubic cubits, which makes the Egyptian estimate only 0.6% in excess of the actual capacity of the granary.

One may ask, then, how the Egyptians managed to arrive at this remarkable result. What made them think of erecting a square on eight-ninths of the circle's diameter, rather than some other amount? It is reasonable to suppose that they might have proceeded in the other direction, starting with the square and dividing its sides into equal segments, three, four or five, say, per side, so that a circle drawn with its centre at the square's centre intersected its sides at points a third, quarter or fifth of the way along from the corners. With the sides of the square divided into thirds, it would immediately be apparent by eye that a circle drawn through all the dividing points would be too small. Using Pythagoras' theorem, it is possible to calculate by how much. If the side of the square was 16 units long, the length of the segments into which the sides were divided would be $5\frac{1}{3}$ units. The length of a half side would be 8 units and the length of a half segment would be $2\frac{2}{3}$ units, and by Pythagoras' theorem the square of the radius of a circle drawn through the dividing points would be equal to $8^2 + (2\frac{2}{3})^2 = 64 + 7\frac{1}{9} = 71\frac{1}{9}$. The area of such a circle would be $71\frac{1}{9}\pi = 223.4$ square units. The area of the square is $16^2 = 256$ square units. It follows that the area of the circle is short by 32.6 square units, which is 12.7%.

Suppose now the scribe, having failed to get a good area match by dividing the side into thirds, next tried divisions into quarters (fig. 9). If he then drew a circle through all the points a quarter of the way from the corners, he would at once see that it had roughly the same area as the square, that is to say, the amount of circle outside the

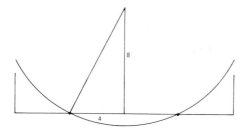

Fig.9. *Divisions into quarters.*

square was about the same as the amount of square outside the circle. Again, it is possible to calculate how well the areas match. The length of a half-side of the square is still 8 units, but the square of the radius of the circle is now $8^2 + 4^2 = 64 + 16 = 80$. The area of the circle is $80\pi = 251.3$ square units, falling short of the area of the square by only 4.7 square units, which is a deficit of 1.8%.

If the end of the radius of the circle had been taken a fifth of the way along the side of the square instead of a quarter, the area of the circle would have been 273.4 square units, giving an excess of 6.8% over the area of the square.

It is reasonable to conclude, therefore, that the Egyptians may initially have 'squared' the circle by drawing it through the quarter-divisions of the sides of the square. However, this does not yet explain their formula. To account for it, a possible clue may be found in fig. 9. The length of the radius is equal to $\sqrt{80} = 8.944$ units, which is short of 9 by only 0.056 units or 0.6%. If it were taken to be 9 units, then the rule adopted by the Egyptians would hold. We suggest that they may have drawn a diameter of the circle through the centre of the square as in fig. 10. If the square is considered now to have a side of 8 units, the diameter of the circle is found by measurement to be approximately 9 units; it may even have been known that a triangle with sides of 4, 8 and 9 units is nearly right-angled. It is significant that the circular base in RMP no. 41 has a diameter of 9 units. The true area of a circle with this diameter will be $\pi \times \frac{1}{4} \times 9^2 = 63.6$ square units. The area as calculated by the Egyptian method is 64 square units: an excess of 0.4 square units or 0.6%. It turns out, then, that the calculation actually gives a better result than was obtained by constructing a square such that the circle intersects quarter points along its sides.

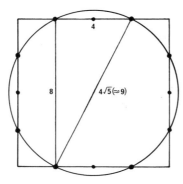

Fig.10. *Squaring the circle.*

The Egyptian method gives the equivalent of a value for π of 3.16 instead of its true value of 3.1416. It is sad that this excellent approximation afterwards fell out of use. In a third century BC demotic papyrus now in Cairo museum, the rule used to calculate the diameter of a circle from its area gives the diameter as equal to the square root of one and a third times the area. This is equivalent to a value of 3 for π, which is out by 4.5%. A similar value of 3 also follows from the Babylonian method of calculating the area of a circle by taking one-twelfth of the square of the circumference.

In RMP no. 42, the problem is to work out the volume of a cylindrical granary, as in no. 41, but as the diameter of the circular base is 10 cubits instead of 9, the calculation is more complicated. In the course of it, $8 + \bar{\bar{3}} + \bar{6} + \overline{18}$ has to be squared. The result of the squaring is given as $79 + \overline{108} + \overline{324}$, although this could have been reduced to $79 + \overline{81}$. The answers to nos. 41 and 42 were given initially in cubic cubits and afterwards converted first to *khar* and then to hundreds of quadruple *hekat*. RMP no. 43, unfortunately marred by a number of errors, was intended to demonstrate a simplified method for finding the volume directly in *khar* without first going through the stage of calculating the volume in cubic cubits. The rule, which gives an exact result, is to add to the diameter its third part, square, and multiply by two-thirds of the height. This formula depends on the identity, expressed in modern terms with d = diameter, h = height:

$$(\tfrac{8}{9})^2 d^2 h \text{ cubic cubits} = \tfrac{3}{2} \times (\tfrac{8}{9})^2 d^2 h \ khar$$
$$= \tfrac{32}{27} d^2 h = \tfrac{2}{3} \times (\tfrac{4}{3})^2 d^2 h \ khar$$

In Egyptian terms, therefore, the scribe who devised the method had to work out the following connection to obtain the volume in *khar* from the diameter and height:

$$(\bar{2} \times 3)(\bar{9} \times 8)(\bar{9} \times 8) = \overline{27} \times 32 = (\bar{3} \times 2)(\bar{3} \times 4)(\bar{3} \times 4)$$

The units of measurement used in RMP nos. 48 and 50 show that these problems are concerned with circular areas of land. No. 48, which is accompanied by a rough diagram, requires the area of a square with a side of 9 *khet* to be compared with that of a circle with a diameter of 9 *khet* inscribed in it. The area of the square is shown to be 8,000 cubit strips + 1 *setat* (= 81 *setat*) and that of the circle to be 6,000 cubit strips + 4 *setat* (= 64 *setat*).

RECTANGLES, TRIANGLES AND PYRAMIDS: RMP nos. 44–46, 49, 51–60

The previous section showed that the Egyptians calculated the volume of cylindrical grain stores correctly by multiplying the area of the base by the height. The volumes of cubical and rectangular containers are determined in the same way in RMP nos. 44–46. Nos. 49, 51 and 52 deal with the areas of rectangular and triangular pieces of land. The rough sketches illustrating the latter have the triangles on their sides with the base to the left, not upright as it would be natural to draw them today. In no. 51, in spite of past arguments concerning interpretation, it is now agreed that the area of a triangle is correctly taken as half that of a rectangle on the same base and with the same height. In no. 52 the area of a truncated triangle is obtained by multiplying the mean of the base and cut side by the height. The method is not illustrated, but the validity of the procedure is apparent from fig. 11.

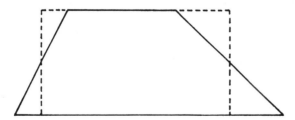

Fig. 11. *Area of a truncated triangle.*

RMP no. 51 can be seen as a special case of no. 52 where the length of the cut side is reduced to zero. No. 53 is also concerned with triangles but unfortunately, through faulty copying, is now incomprehensible. Nos. 54 and 55 deal with areas of land, but are not concerned with specific shapes.

RMP nos. 56–60 are about architectural structures with sloping sides. Except in the case of the last problem, these are pyramids. The object is to show how the slope of the faces relates to overall dimensions. The unit of slope, as explained above in the section 'Numerals and units of measurement' (page 15), is the *seked*, defined as the lateral displacement in palms for a drop of 7 palms (= 1 royal cubit). This is the nearest the RMP gets to trigonometry. In the RMP the *seked* is taken to be half the width of the base divided by the height, both expressed in cubits, and then multiplied by 7.

All the problems are illustrated by rough sketches; these do not accurately represent the actual *seked*, which is shown too steep. In problems 57–59, the *seked*, whether given or to be found, is in each case $5 + \overline{4}$ palms, the same as in the pyramid of Chephren. In no. 57 the pyramid has a base of 140 cubits and it is required to find the height. This is obtained by dividing 7 by twice the *seked* to get $\overline{\overline{3}}$, which is then multiplied by 140 to give a height of $93 + \overline{3}$ cubits. In no. 58, the dimensions of the pyramid are the same but the *seked*, not the height, has to be found. Half the base is divided by the height to give $\bar{2} + \bar{4}$ cubits, which is then multiplied by 7 to convert it to palms. The dimensions in no. 59 are much smaller, being those of a miniature pyramid such as was sometimes erected outside a private tomb. The *seked* is found by halving the base, which is 12 cubits, and dividing it by the height, which is 8 cubits, to get $\bar{2} + \bar{4}$ cubits. This is then multiplied by 7 as in no. 58.

In RMP no. 56, it is required to find the *seked* of a pyramid with base 360 cubits and height 250 cubits. The base is halved and divided by the height to get $\bar{2} + \bar{5} + \overline{50}$ cubits. This multiplied by 7 gives a *seked* of $5 + \overline{25}$ palms. It is likely that this problem was intended as an exercise rather than to reflect a real situation, since the evidence from surviving pyramids is that the stone-masons cut the facing stones in quarters of a *seked*, not in twenty-fifths, which would probably not have been possible. The claim sometimes made that the lower part of the Bent Pyramid has a *seked* of $5 + \overline{25}$ is incorrect; the *seked* is, in fact, exactly 5.

In the last of the slope problems, no. 60, which is also the last problem on the recto of BM 10057, the *seked* has to be obtained for a structure of doubtful nature, twice as high as it is wide at the base. The width of the base is 15 cubits while the height is 30 cubits. The scribe halves the base and divides by the height as usual, but unfortunately he, or the copyist, forgot to multiply by 7 to convert to palms. The result for the *seked* should have been $1 + \bar{2} + \bar{4}$, a plausible slope for walls with a steep batter.

It is remarkable that there is no pyramid problem in the RMP dealing with volume, although it would surely have been desirable from a practical point of view to be able to calculate the amount of stone that would be required to build a pyramid of a particular size. The rule for obtaining the volume of a pyramid is to multiply the area of the base by a third of the height. Put another way, the volume of a pyramid is equal to one-third that of a prism on the same base with the same height. This was shown to be true for a pyramid on a triangular base by Euclid; in *Elements* xii, 7 a prism on a triangular base is dissected along three diagonals to form three such pyramids of equal volume. The same formula must hold for a pyramid on a square base, since it can be divided into two equal pyramids on triangular bases by a section, vertical in the case of a right pyramid, passing through the apex and a diagonal of the base.

The question is, did the ancient Egyptians, by whatever means, know the rule for determining the volume of a pyramid? They certainly did by the third century BC, because the demotic text mentioned above in the section 'Squaring the circle' (page 45) contains a pyramid problem in which the volume is found correctly. It seems highly likely that the formula was known earlier, in view of a famous problem in the Moscow Mathematical Papyrus, which, as stated above (page 10), is judged to be a little earlier than the RMP. In this problem, the method is given, correctly, for determining the volume of a truncated pyramid. The volume of a complete pyramid would be a special case of this, where the area of the cut (top) face is reduced to zero.

Since there is no clue as to how the Egyptians achieved their result, it may be worthwhile to consider how it can be arrived at by modern methods. This can be done either by completing the truncated pyramid, or frustum, and subtracting the volume of the smaller, added pyramid from that of the larger, complete one; or by dissecting the frustum into four separate pieces whose shapes are such that their volumes can be calculated and added together.

According to the subtraction method, if the width of the base of the frustum is a and that of the top face is b, then the heights of the two pyramids will be proportional to their base widths and so can be written ka and kb, where k is a common factor (fig. 12).

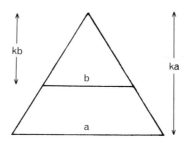

Fig.12. *Side view of a frustum completed to form a pyramid.*

The volume of the frustum is equal to $(\frac{1}{3}ka \times a^2) - (\frac{1}{3}kb \times b^2)$ or $\frac{1}{3}k(a^3 - b^3)$. But $a^3 - b^3$ can be factorised into $(a-b)(a^2 + ab + b^2)$, so that, since $k(a - b) =$ height of frustum $= h$,

volume of frustum $= \frac{1}{3}h(a^2 + ab + b^2)$

which is the formula of the Moscow Mathematical Papyrus.

Figure 13 illustrates how the volume can be determined by dissection. It avoids having to factorise the difference between two cubes. The frustum is divided into: a right pyramid on a square base, of side $a - b$ (top left); an oblique square prism, with base of side b (bottom right); and two equal oblique triangular prisms, each resting on a face with dimensions $b \times (a-b)$, seen at the top right and bottom left of the diagram. All the figures as positioned have an effective altitude h, which is the height of the frustum. The volumes are as follows:

pyramid	$\frac{1}{3}h(a - b)^2$
square prism	$h \times b^2$
two triangular prisms	$h \times (a - b) \times b$

summing to

$\frac{1}{3}h[(a^2 - 2ab + b^2) + 3b^2 + (3ab - 3b^2)]$
$= \frac{1}{3}h(a^2 + ab + b^2)$

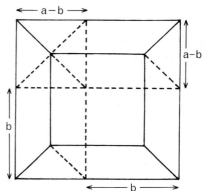

Fig.13. *Frustum of a pyramid viewed from above.*

The earliest proof of the formula for the volume of a truncated pyramid was given by the Greek mathematician Heron of Alexandria, *Metrica* ii, 8. Heron's date is uncertain, but it lay somewhere between 150 BC and AD 250. We leave it to the reader to form his own conclusions as to whether the Egyptians had the necessary expertise to solve the problem by precise mathematics, whether the correct answer was more likely to have been reached empirically — or whether it was, as some have supposed, the outcome of a lucky guess.

VALUE, FAIR EXCHANGE AND FEEDING
RMP nos. 62, 66, 69–78, 82–84

As will have been observed, the RMP contains a mixture of numerical manipulation and practical problem-solving. One of the commonest practical activities of ancient Egypt must have been weighing. Even in the imagined after-life a frequently depicted scene was the weighing, on a magnificent balance, of the deceased's heart against the feather of Truth. It is remarkable, therefore, that only once are weights mentioned in the RMP; this is in a problem, no. 62, concerning a bag that contained equal weights of gold, silver and lead together worth 84 *shaty*. In an economy that operated without coinage, the *shaty* perhaps functioned as an accounting device to indicate value for purposes of exchange. In this problem, the 'price' per *deben*, which was a unit of weight approximately equal to 91 grams, is given for each metal as: gold 12 *shaty*; silver 6 *shaty*; lead 3 *shaty*. It is required to find the 'price' for each metal. The calculation goes as follows:

Total price		= 84 *shaty*
Total price for 1 *deben* of gold, 1 *deben* of silver and 1 *deben* of lead	= 12 + 6 + 3	= 21 *shaty*
Weight of each metal	= 84 ÷ 21	= 4 *deben*
Price of the gold	= 12 × 4	= 48 *shaty*
Price of the silver	= 6 × 4	= 24 *shaty*
Price of the lead	= 3 × 4	= 12 *shaty*
	Total	= 84 *shaty*

Fig.14. *Recording the weighing of gold, from the tomb of Rekhmire at Thebes.*

Note that gold is rated as being twice as valuable as silver, in spite of the abundant supply of gold in Egypt and the need to bring in silver from outside sources.

A food problem that is highly practical is RMP no. 66, which asks: if 10 *hekat* of fat are issued to last a year, how much can be used up in a day? The Egyptian year, like ours, consisted of 365 days, a number that could not be adjusted to produce a neat answer to the problem. To solve it, 10 *hekat* were converted to *ro*, and the result was ponderously divided by 365 to give $8 + \bar{\bar{3}} + \overline{10} + \overline{2190}$ *ro*, which, expressed in Horus-eye fractions, since $\overline{64}$ *hekat* = 5 *ro*, is $\overline{64}$ *hekat* $+ 3 + \bar{\bar{3}} + \overline{10} + \overline{2190}$ *ro*. The division of 3200 by 365 could have been simplified to $640 \div 73$. Using the partial products and remainder method described above in the section 'Multiplication and division' (page 18), it can be written:

	Partial products	Remainders
73×8	584	$640 - 584 = 56$
$73 \times \bar{\bar{3}}$	$48 + \bar{\bar{3}}$	$56 - (48 + \bar{\bar{3}}) = 7 + \bar{\bar{3}}$
$73 \times \overline{10}$	$7 + \bar{5} + \overline{10}$	$(7 + \bar{\bar{3}}) - (7 + \bar{5} + \overline{10}) = \overline{30}$
$73 \times \overline{2190}$	$\overline{30}$	nil

If food is to be exchanged without a money transaction, it is necessary to have a measure of quality. Such exchange could have been a frequent occurrence, so it is not surprising that as many as ten problems on the verso of BM 10058, nos. 69–78, are devoted to the matter. The two commodities that were the staples of the ordinary man's diet were bread and beer, and the common factor to which value could be attached apart from labour, which was presumably expendable, was the grain from which they were made. The bread might be more or less aerated or otherwise expanded, and the beer more or less dilute; so it would not be adequate in exchanging bread for beer, which was the subject of nos. 77 and 78, to trade so many loaves of the former for so many jugs of the latter, each of a certain size. Hence the *pesu* unit, which measures lack of quality by indicating the number of loaves or jugs of beer that were obtained from 1 *hekat* of flour or grain.

Fig.15. *Brewing beer from bread, from the mastaba of Khentika called Ikhekhi at Saqqara.*

Since the mathematics of the *pesu* problems is in the main straightforward, we shall confine ourselves to two examples, nos. 71 and 72, the first of which deals specifically with the dilution of beer, while the second, dealing with the exchange of bread, has opted, presumably deliberately, for a roundabout solution. In no. 71, a quarter of the contents of a jug of beer is poured off and replaced with water. It is required to discover the new *pesu*, knowing that the original jugful was the product of half a *hekat* of grain. A quarter of this, namely $\bar{8}$ *hekat*, is subtracted to give $\bar{4} + \bar{8}$ *hekat*, which is then divided into 1, there being one jug, to give a *pesu* of $2 + \bar{\bar{3}}$. This is another example of a problem being solved by obtaining a reciprocal; it is not explained how the reciprocal was got, but a simple division sum would suffice:

$$1 \qquad\qquad \bar{4} + \bar{8}$$
$$/2 \qquad\qquad \bar{2} + \bar{4}$$
$$/\bar{\bar{3}} \qquad\qquad \bar{6} + \bar{12}$$

$$\text{Total } 2 + \bar{\bar{3}} \qquad\qquad \bar{2} + \bar{4} + \bar{6} + \bar{12}$$
$$= 1, \text{ since } \bar{6} + \bar{12} = \bar{4}$$

In RMP no. 72, it is asked how many loaves of *pesu* 45 should be exchanged for 100 loaves of *pesu* 10. The straightforward solution to this problem, as adopted in other *pesu* problems, would consist of the following steps:

100 loaves of *pesu* 10 are made from $100 \div 10 = 10$ *hekat* of flour;
10 *hekat* of flour will make $10 \times 45 = 450$ loaves of *pesu* 45.

The scribe, however, in this instance chose a more complicated approach via the excess in the *pesu*:

excess *pesu* $= 45 - 10 = 35$;
pesu increased by a factor of $35 \div 10 = 3 + \bar{2}$ to produce this excess;
number of loaves multiplied by the same factor gives an excess of $100(3 + \bar{2}) = 350$;
total number of loaves to be exchanged $= 350 + 100 = 450$.

In brief, the calculation is $[100 \times \overline{10}(45 - 10)] + 100 = 450$; or, in general terms, the answer is got from $y = [(b\text{-}a)/a]x + x$ instead of $y = (b/a)x$. There can be no advantage in this procedure, except to demonstrate an alternative method of tackling the problem.

Fig.16. *Force-feeding cranes, from the mastaba of Ti at Saqqara.*

The last true problems in the RMP, nos 82–84, are concerned with the feeding of poultry and cattle. Their main interest lies in the information that they provide on the amount of fodder supplied to different animals under different conditions. Problem 82 gives the example of 10 geese being fattened by force-feeding and receiving between them $2 + \bar{2}$ *hekat* of flour, made into bread, per day. In no 82B the amount is half this.

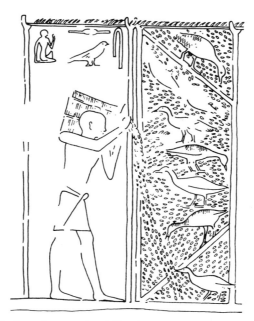

Fig.17. *Feeding free-range birds, from the mastaba of Ti at Saqqara.*

According to no. 83 the same number of birds, if not getting special treatment, would receive only 1 *hekat* of grain, and birds in coops a quarter of this, presumably because of their reduced activity. In order to make a true comparison, it is necessary to allow for an increase in bulk that occurs when grain is ground to make flour. The RMP offers a rule of thumb for determining the increase. To discover the amount of grain in *hekat* required to yield 100 *hekat* of flour, take two-thirds of 100 and one-tenth of that, and subtract from 100:

$$100 \times \bar{\bar{3}} = 66 + \bar{\bar{3}}$$
$$(66 + \bar{\bar{3}}) \times \overline{10} = 6 + \bar{\bar{3}}$$
$$100 - (6 + \bar{\bar{3}}) = 93 + \bar{3}$$

Since $\bar{\bar{3}} \times \overline{10} = \overline{15}$, the procedure is equivalent to reducing the amount of flour by a fifteenth part to discover the amount of grain needed to produce it. The amount of flour force-fed daily to 10 geese, namely $2 + \bar{2}$ *hekat*, would have been ground from $(2 + \bar{2}) - \overline{15}(2 + \bar{2}) = 2 + \bar{2} - (\overline{10} + \overline{15})$ *hekat* of grain, since $\overline{15} \times 2 = \overline{10} + \overline{30}$. Because $\bar{3} + \overline{10} + \overline{15} = \bar{2}$, it follows that $\bar{2} - (\overline{10} + \overline{15}) = \bar{3}$. Therefore, the force-fed geese are given $(2 + \bar{3}) \times$ as much feed as free-range birds.

As beer was commonly brewed from fermented bread rather than directly from grain, in most of the *pesu* problems beer and bread are both assessed in terms of flour content, so that the increase in bulk that occurs during grinding does not affect the exchange. In problem no. 71, where beer is simply diluted and not exchanged for bread, the *pesu* is assessed in terms of grain content and is, therefore, not comparable with the *pesu* of the other problems involving beer. A jugful of beer made from a particular volume of grain must be stronger than a jugful made from the same volume of flour. In one problem, no. 74, dealing with bread only, the increase in volume that occurs in milling grain is explicitly ignored.

In view of Plato's account, given on page 4, of the teaching of mathematics to Egyptian children through play, one might expect to find an element of fun creeping into the RMP. However, Plato's remarks relate to instruction given at an early age, whereas the RMP, which is on the whole a solemn document, can be seen as a teacher's manual to be used in the training of the more mature. Nevertheless, there are three problems that could conceivably have a lighter side. Two of them, nos. 28 and 29, appear among the 'unknown quantity' problems. They have the general form of first-degree equations requiring solution like the other members of the group, but Gillings has suggested that they might be archetypal 'Think of a number' problems, in which a member of the audience performs a calculation on a number chosen by himself without the scribe's knowledge. He then announces the result, whereupon the scribe does his own calculation and is able to come up with the original number.

In RMP no. 28 the equation has the form:

$$x + \bar{\bar{3}}x - \bar{3}(x + \bar{\bar{3}}x) = 10$$

On the 'Think of a number' interpretation, the 'victim' is asked to add to his chosen number two-thirds of itself, to take away a third of the result, and to give the answer. If the answer is 10, the scribe is able to say that the number first thought of was 9. This he does by simply subtracting one-tenth of the answer from itself. The RMP working gives this method of solving the equation from the calculation $10 - (10 \times \overline{10}) = 9$, and proves that 9 is the correct solution by the following steps:

$$9 \times \bar{\bar{3}} = 6$$
$$9 + 6 = 15$$
$$15 \times \bar{3} = 5$$
$$15 - 5 = 10$$

The proof demonstrates that 9 satisfies the terms of the problem.

The objection to the Gillings hypothesis is that great difficulties might be encountered if the number chosen was not a multiple of 3. Supposing the victim has selected 13 rather than 9, his calculation would have to go through the following stages:

$$13 \times \bar{\bar{3}} = 8 + \bar{\bar{3}}$$
$$13 + (8 + \bar{\bar{3}}) = 21 + \bar{\bar{3}}$$
$$(21 + \bar{\bar{3}}) \times \bar{3} = 7 + \bar{6} + \overline{18}$$
$$(21 + \bar{\bar{3}}) - (7 + \bar{6} + \overline{18}) = 14 + \bar{\bar{3}} - (\bar{6} + \overline{18})$$
$$= 14 + \bar{\bar{3}} + \bar{9}, \text{ since } \bar{6} + \bar{9} + \overline{18} = \bar{3}$$

The scribe would then have to subtract a tenth part from $14 + \bar{3} + \bar{9}$, no easy task without it first being transformed into $\bar{9} \times 130$. If he had the leisure to do this, the calculation might continue:

$$(\bar{9} \times 130) \times \overline{10} = \bar{9} \times 13$$
$$(\bar{9} \times 130) - (\bar{9} \times 13) = \bar{9} \times 13 \times (10 - 1)$$
$$= 13$$

In view of the complexities that were liable to be encountered, it is safe to say that RMP no. 28 could have acted as a 'Think of a number' game only with the proviso that the number thought of must be a multiple of 3.

The proper view to take of RMP no 28 is probably that the equation is merely an elaborate version of $\bar{\bar{3}}(x + \bar{\bar{3}}x) = 10$ which, since $\bar{\bar{3}}$ is the reciprocal of $1 + \bar{2}$, can be further simplified to $(1 + \bar{\bar{3}})x = 15$. In this form it would have been analogous to RMP nos. 24–27 immediately preceding, where the equations to be solved are $(1 + \bar{7})x = 19$, $(1 + \bar{2})x = 16$, $(1 + \bar{4})x = 15$ and $(1 + \bar{5})x = 21$. $(1 + \bar{\bar{3}})x = 15$ is easily solved by partitioning 15 into $9 + 6$, which is equal to $9(1 + \bar{\bar{3}})$. It is then obvious that $x = 9$.

Having expanded the basic equation as part of a policy to produce problems of increasing difficulty, the scribe could have worked on the coefficient of the unknown in $[(1 + \bar{\bar{3}}) - \bar{\bar{3}}(1 + \bar{\bar{3}})]x = 10$ as follows:

$$(1 + \bar{\bar{3}}) - \bar{\bar{3}}(1 + \bar{\bar{3}}) = 1 + \bar{3} - (\bar{9} \times 2)$$
$$= 1 + \bar{3} - (\bar{6} + \overline{18})$$
$$= 1 + \bar{9}$$
$$= \bar{9} \times 10$$

Since the reciprocal of $\bar{9} \times 10$ is $\overline{10} \times 9 = 1 - \overline{10}$, it follows that the unknown is equal to the number 10 less a tenth, and if $x + \bar{\bar{3}}x - \bar{\bar{3}}(x + \bar{\bar{3}}x)$ had been equated to any other number rather than 10 the same would have been true — that is, the answer could be obtained by subtracting a tenth part from that number.

In RMP no. 29, the beginning is missing, having been omitted by a copying error, but the problem has been reconstructed to mean: Find a number such that, when two-thirds of it is added and then one-third of the sum, a third of the total is equal to 10. In other words, it is required to solve:

$$\bar{3}[x + \bar{\bar{3}}x + \bar{\bar{3}}(x + \bar{\bar{3}}x)] = 10$$

The scribe gets the answer by adding a quarter and a tenth of 10 to 10, to obtain $13 + \bar{2}$. He then shows that $13 + \bar{2}$ satisfies the conditions of the problem.

For the same reasons as in RMP no. 28, no. 29 can function as a 'Think of a number' game only if the number thought of is a multiple of 3. The equation can be rewritten as:

$$\bar{3}(1 + \bar{3})(1 + \bar{\bar{3}})x = 10$$

and this can be simplified since $10 \div (1 + \bar{\bar{3}}) = 6$. The reduced form, therefore, is $(1 + \bar{3})x = 18$, again analogous to RMP nos. 24–27. The solution to this equation is easily shown to be $13 + \bar{2}$.

The scribe's method for obtaining a solution can be arrived at by manipulating the coefficient of x as follows:

$$\overline{3}(1 + \overline{3})(1 + \overline{3}) = \overline{3}[2 + (\overline{9} \times 2)]$$
$$= \overline{3}(1 + \overline{9})$$
$$= \overline{3} \times \overline{9} \times 10$$
$$= \overline{27} \times 20$$

Using the same procedure as in the analysis of RMP no. 28, the reciprocal of $\overline{27} \times 20$ is $\overline{20} \times 27$, and 27 can be partitioned into $20 + 5 + 2$, so that $\overline{20} \times 27 = \overline{20}(20 + 5 + 2) = 1 + \overline{4} + \overline{10}$. It follows that equations of the type $\overline{3}(1 + \overline{3})(1 + \overline{3})x = N$ can be solved by adding to N its quarter and its tenth.

The last problem to be considered, RMP no. 79, is of a very different sort. It can be regarded as a precursor of the well-known nursery catch 'As I was going to St Ives, I met a man with seven wives, Each wife had seven . . .' etc. The Egyptian version is set out as follows:

7	houses
49	cats
343	mice
2401	[ears of] spelt [erroneously written 2301]
16807	*hekat* [of grain]

Total 19607

The numbers form five terms of a divergent geometrical progression, of which the first term, 7, is the same as the common ratio. It is intended to show that in each of 7 houses there are 7 cats, each cat catches 7 mice, each mouse would eat 7 ears of corn, each ear of corn would, if sown, produce 7 *hekat* of grain.

The mathematical interest of the problem lies in the revelation that the Egyptians understood how to sum the terms of a progression of this type. The following working is provided in a separate column:

1	2801
2	5602
4	11204

Total	19607

This can be recognised at once as a sum in which the number 2801 is multiplied by 7 to get 19607, which was the total obtained before by adding the numbers of houses, cats, mice, etc. The significance of the 2801 becomes apparent when the original series is summed term by term:

7	7
7×7	$7(1 + 7) = 56$
$7 \times 7 \times 7$	$7(1 + 56) = 399$
$7 \times 7 \times 7 \times 7$	$7(1 + 399) = 2800$
$7 \times 7 \times 7 \times 7 \times 7$	$7(1 + 2800) = 19607$

It will be seen that at each stage the sum is obtained by increasing the previous sum by 1 and multiplying by the common ratio. The sum of the series to four terms was 2800. Increased by 1 this becomes 2801. Multiplying 2801 by 7 gives the sum to five terms, and this was what the scribe did.

CONCLUSIONS

In our preface we applied to the RMP the epithet 'great'. Its greatness, if that accolade is justified, lies not so much in its method of presenting material as in the glimpses that can be obtained through it of man's burgeoning intellectual powers. The RMP is the most extensive mathematical source-book that survives from ancient Egypt, but it must be remembered that Egyptian attainments may have been greater than its pages reveal; one cannot expect the full range of knowledge to be apparent in a learner's text. How far back does that knowledge go? On the evidence of the copyist Ahmose, the RMP reproduces material current in the Middle Kingdom, but its origins must date back to Old Kingdom days, at least to the age of the stone pyramid-builders if not before. Surviving pyramids indicate that the *seked* of $5\frac{1}{4}$ palms featuring in RMP nos. 57–59 was brought into use during the Fourth Dynasty, became universal in the Sixth, and had been abandoned by the Twelfth. One may suppose, therefore, that these particular problems were devised somewhere between the twenty-second and twenty-fifth centuries BC, and that one is looking into the minds of a people who achieved marvellous work over 4,000 years ago. The Egyptian civilisation is remarkable for its early flowering, and the mathematics, like the hieroglyphic script and the conventions of its art, may hark back to the very beginnings.

The practising mathematicians were the scribes. Even before the time of Ahmose they regarded themselves as forming an elite. The Middle Kingdom author of *The Satire of Trades* lauded the profession of scribe as the greatest of callings, to be loved more than one's mother. In the New Kingdom, a scribe called Hori gave some fascinating revelations of professional self-regard in a *Satirical Letter* (Papyrus Anastasi I, BM 10247), in which he castigates a colleague for his inadequacies, at the same time pointing out the sort of expertise that a good scribe should possess. This included the ability to organise the digging of a lake, the building of a ramp, the transport of an obelisk, the erection of a colossus, the provisioning of a military mission, and even a knowledge of Asiatic geography. The scribal activities portrayed in the RMP are more run-of-the-mill, possibly because the apprentice must walk before he can run. Nevertheless, the training provided would have sufficed for most professional purposes, judging by the content of surviving administrative documents. One of the most important assignments was the keeping of accounts for the numerous institutions of ancient Egypt; in tomb scenes, almost every activity is shown with attendant scribes busily recording what is going on.

Although Egyptian arithmetic clearly had a strong practical slant and must have had its origins in the needs of society, those very needs required an agility in the manipulation of numbers that may have engendered an interest in the properties of numbers for their own sake. Nowhere was greater skill and versatility shown than in the deployment of unit fractions. This was at once the glory and the straitjacket of Egyptian

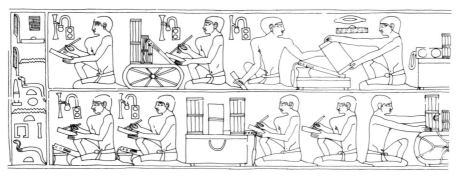

Fig.18. *Accounting by the assessors of the funerary estate, from the mastaba of Ti at Saqqara.*

methodology; some of the usages persisted into Greek and even Roman times. The Greeks often proclaimed their mathematical debt to Egypt. Proclus in his *Commentary on Euclid*, following Herodotus, wrote that geometry had an Egyptian origin arising out of the necessity of resurveying the land after each inundation. Aristotle (*Metaphysics* I, i, 16), on the other hand, attributed the birth of mathematics in Egypt to the existence of a priestly leisured class. Perhaps the most enduring effect of Egyptian mathematics was the stimulus that it gave to the Greeks, who then travelled beyond mere calculation into the realms of abstract thought. In which case, the Greek achievement will have owed something to the learning meticulously and patiently handed on by the Egyptians from generation to generation since early times.

SELECT BIBLIOGRAPHY

Bruins, E.M. 'Ancient Egyptian arithmetic: 2/n.' *Indagationes mathematicae* **14** (1952), 81–91.

Bruins, E.M. 'Platon et la table égyptienne 2/n.' *Janus* **46** (1957), 253–63.

Černý, J. *Paper and Books in Ancient Egypt*, 1952.

Chace, A.B., Bull, L., Manning, H.P., and Archibald, R.C. *The Rhind Mathematical Papyrus*, Mathematical Association of America, vol. 1 1927, vol. 2 1929, reprinted 1979.

Engels, H. 'Quadrature of the circle in ancient Egypt.' *Historia Mathematica* **4** (1977), 137–40.

Gardiner, A.H. *Literary Texts of the New Kingdom: Anastasi I, A Satirical Letter*, 1911.

Gillings, R.J. *Mathematics in the Time of the Pharaohs*, 1972, reprinted 1982 (reviewed by Bruckheimer, M. and Saloman, Y. *Historia Mathematica* **4** (1977), 445–52).

Gillings, R.J. 'The mathematics of ancient Egypt', in *Dictionary of Scientific Biography*, ed. C.C. Gillespie, 15 suppl. 1, 1978.

Gillings, R.J. 'The recto of the Rhind Mathematical Papyrus. How did the ancient Egyptian scribe prepare it?' *Archive for History of Exact Sciences* **12** (1974), 291–8.

Glanville, S.R.K. 'The Mathematical Leather Roll in the British Museum', *Journal of Egyptian Archaeology* **13** (1927), 232–8.

Heath, T. *A History of Greek Mathematics*, vols. 1, 2, 1921, reprinted 1981.

Hultsch, F. *Die Elemente der Ägyptischen Theilungsrechnung 8, Übersicht über die Lehre von den Zerlegungen*, 1895, 169–71.

Parker, R.A. *Demotic Mathematical Papyri*, 1972.

Peet, T.E. *The Rhind Mathematical Papyrus BM 10057 and 10058*, 1923, (reviewed by Gunn, B., *Journal of Egyptian Archaeology* **12** (1926), 123–37).

Peet, T.E. 'Mathematics in ancient Egypt.' *Bulletin of the John Rylands Library* **15** (1931), 409–41.

Robins, G. 'Mathematics, astronomy, and calendars in Pharaonic Egypt', in *Civilizations of the Ancient Near East*, ed. Jack M. Sasson, III, 1995, 1799–1813.

van der Waerden, B.L. 'The (2:n) table in the Rhind Papyrus'. *Centaurus* **23** (1980), 259–74.

For further reading see the bibliographies of Chace, *op. cit.*, and Gillings, *op. cit.*

PLATES

PLATE 2 DOUBLED FRACTIONS 5̄, 1̄7̄–2̄7̄; ENDS OF 3̄, 7̄–1̄5̄

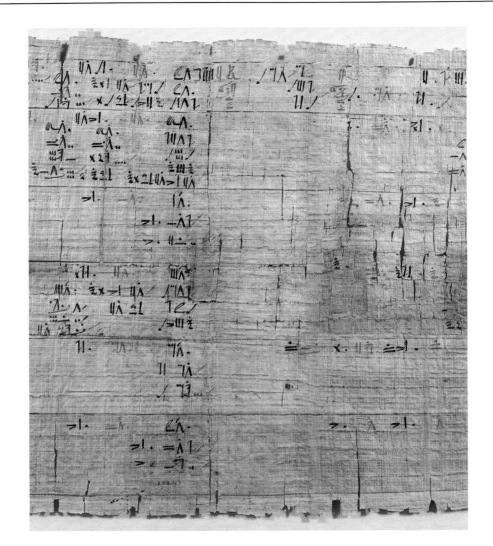

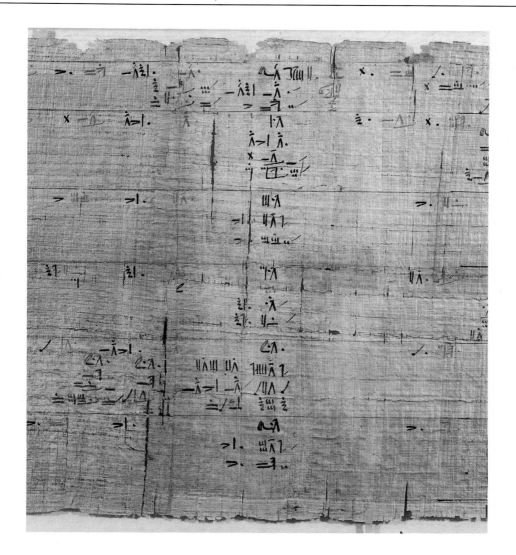

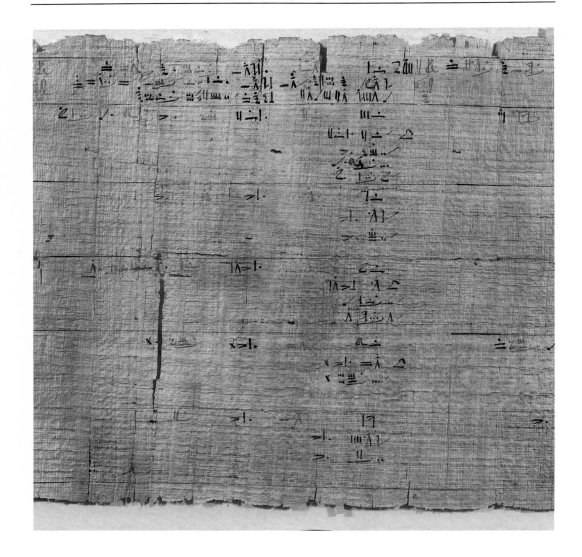

PLATE 5 DOUBLED FRACTIONS 53–65

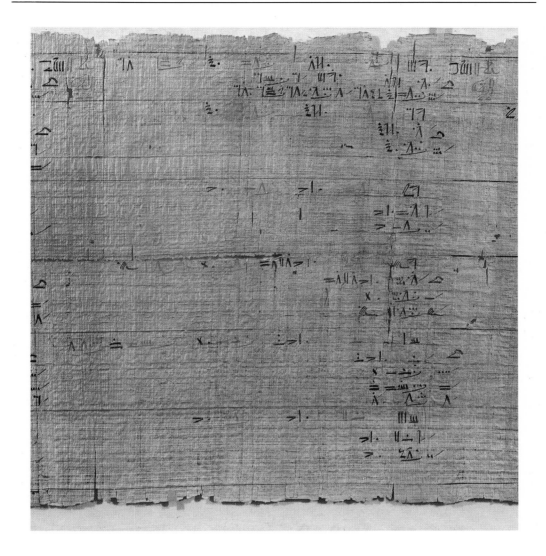

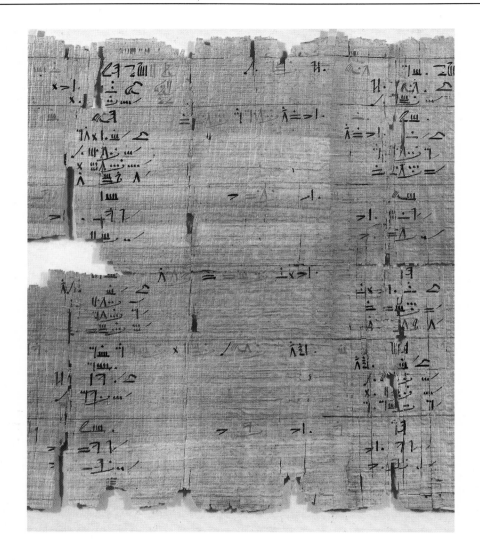

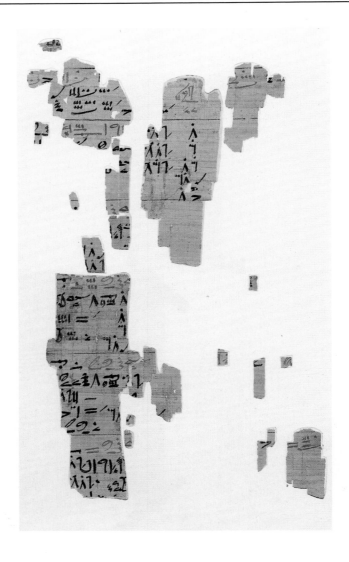

PLATE 9   PROBLEMS 6—23; ENDS OF 1—5

PLATE 10   PROBLEMS 24–30; ENDS OF 21–3

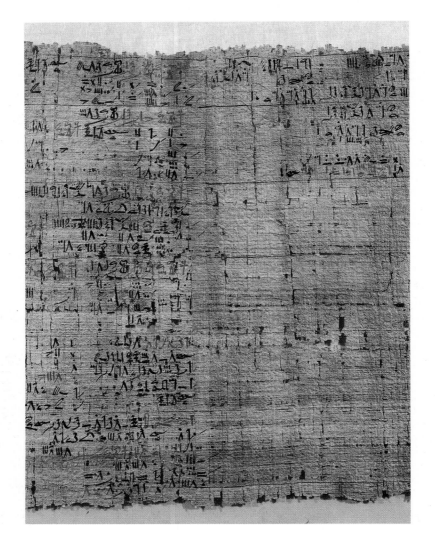

PLATE 11   PROBLEMS 31–3; ENDS OF 24–30

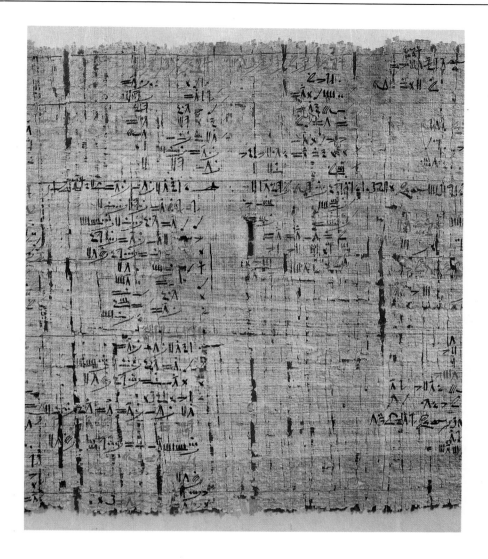

PLATE 12   PROBLEMS 34–8; END OF 33

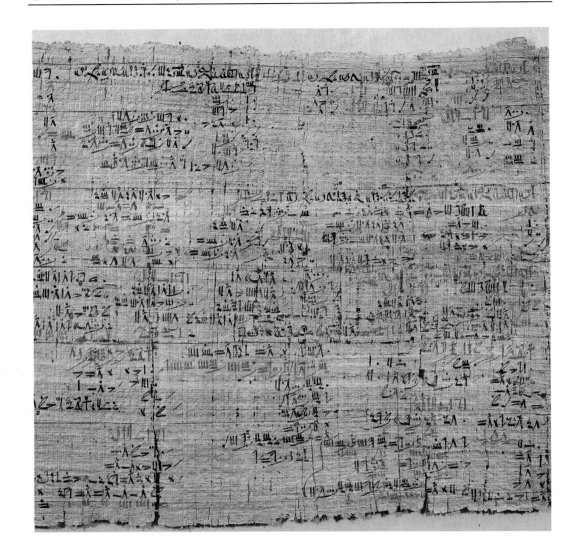

PLATE 13   PROBLEMS 39–40; ENDS OF 34, 36–8

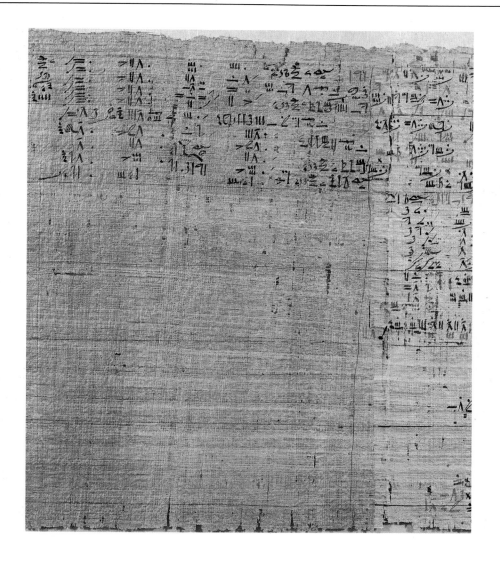

PLATE 14   PROBLEMS 41–6

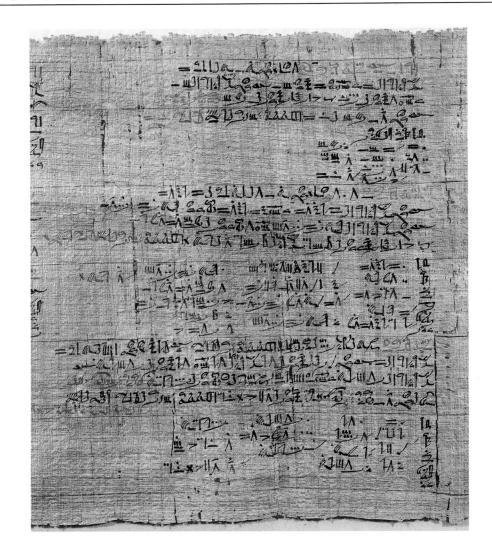

PLATE 15   PROBLEMS 47–8; ENDS OF 43–6

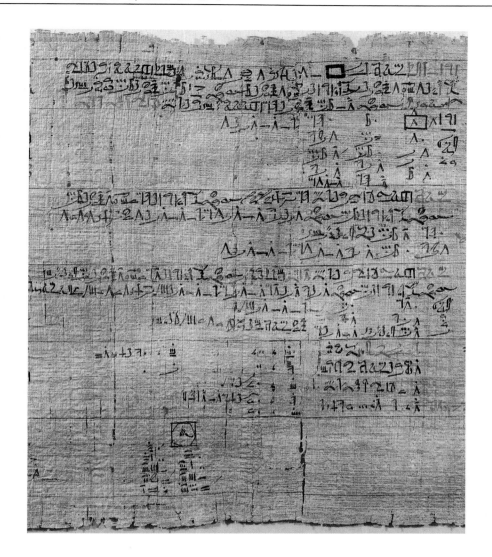

PLATE 16   PROBLEMS 49–55; END OF 46

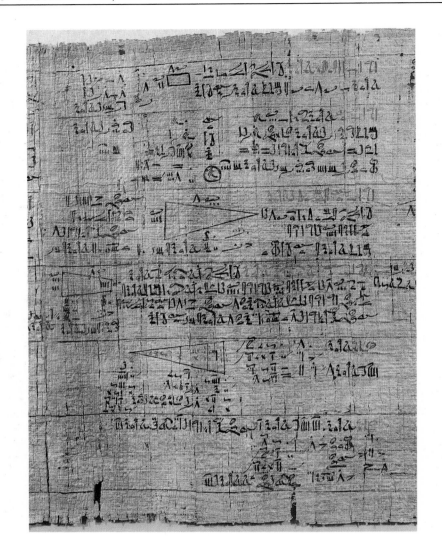

PLATE 17   PROBLEMS 56–60; END OF 52

PLATE 18   PROBLEMS 61–4 (BEGINNING OF *VERSO*)

PLATE 19   PROBLEMS 65–70; END OF 64

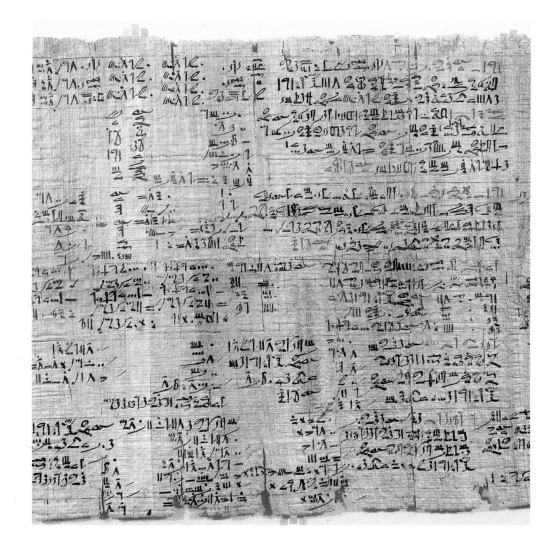

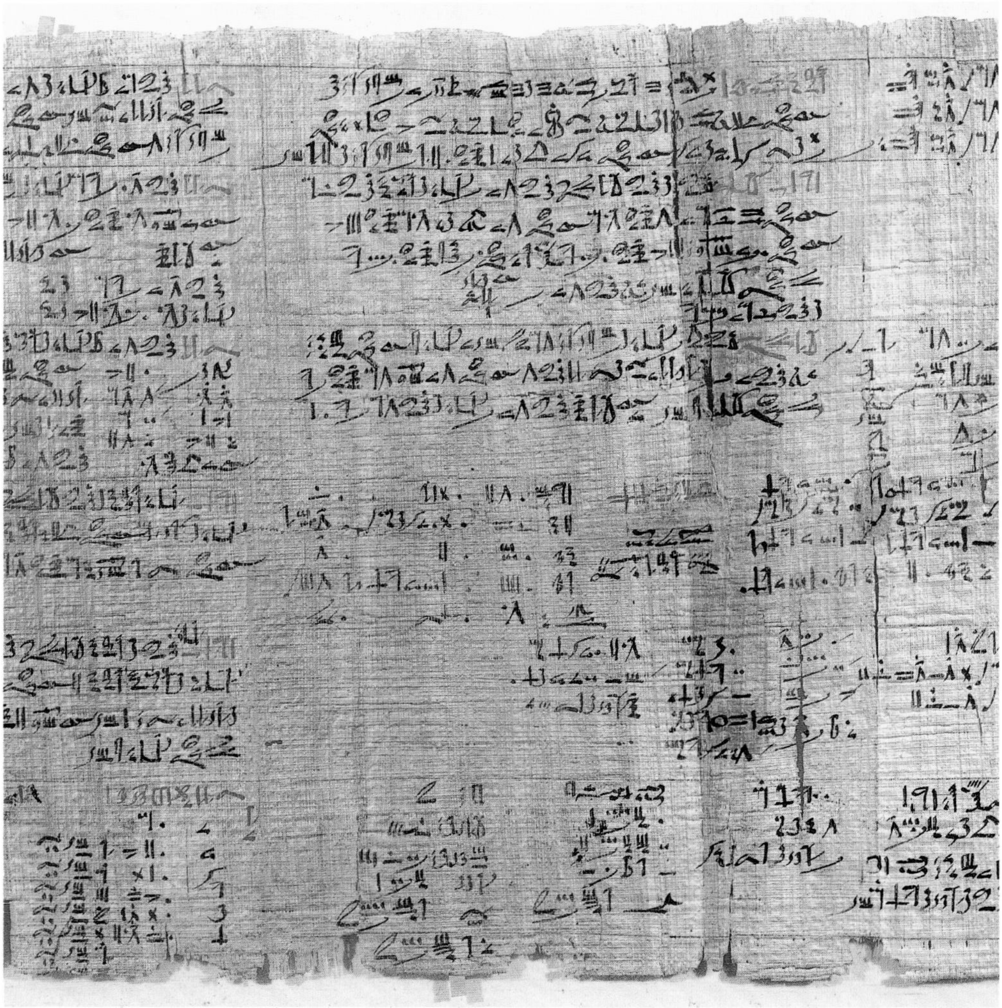

PLATE 20   PROBLEMS 71–9, 81; ENDS OF 65, 67–70

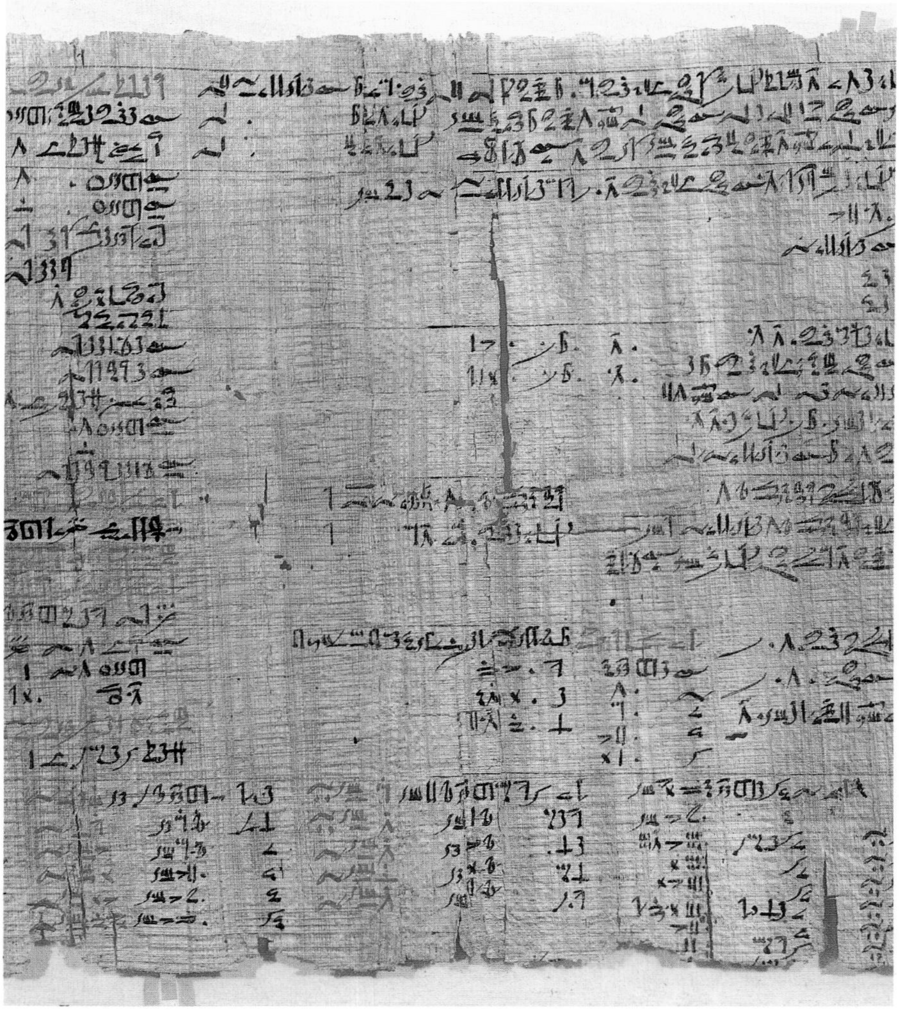

PLATE 21 PROBLEMS 80, 82–3; ENDS OF 74–8

PLATE 22   PROBLEM 84; ENDS OF 81–3

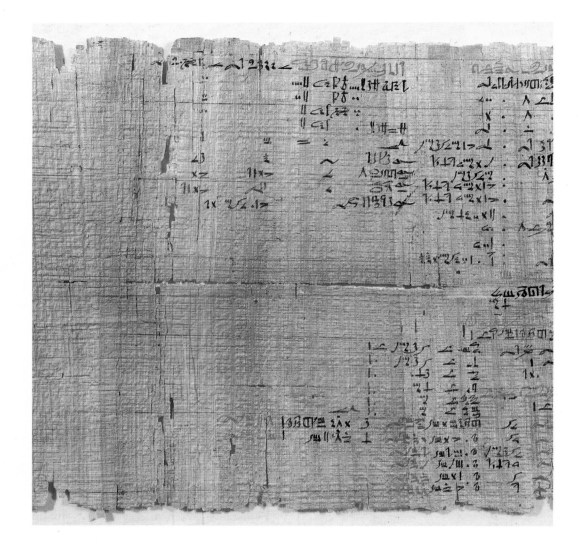

PLATE 23   (ABOVE) NUMBER 85; (BELOW) NUMBER 87

PLATE 24   NUMBER 86